Shock Wave and High Pressure Phenomena

More information about this series at http://www.springer.com/series/1774

Shock Wave and High Pressure Phenomena

L.L. Altgilbers, M.D.J. Brown, I. Grishnaev, B.M. Novac, I.R. Smith, I. Tkach, and Y. Tkach: Magnetocumulative Generators

T. Antoun, D.R. Curran, G.I. Kanel, S.V. Razorenov, and A.V. Utkin: Spall Fracture

J. Asay and M. Shahinpoor (Eds.): High-Pressure Shock Compression of Solids

S.S. Batsanov: Effects of Explosion on Materials: Modification and Synthesis Under High-Pressure Shock Compression

G. Ben-Dor: Shock Wave Reflection Phenomena

L.C. Chhabildas, L. Davison, and Y. Horie (Eds.): High-Pressure Shock Compression of Solids VIII

L. Davison: Fundamentals of Shock Wave Propagation in Solids

L. Davison, Y. Horie, and T. Sekine (Eds.): High-Pressure Shock Compression of Solids

V.L. Davison and M. Shahinpoor (Eds.): High-Pressure Shock Compression of Solids III

R.P. Drake: High-Energy-Density Physics

A.N. Dremin: Toward Detonation Theory

J.W. Forbes: Shock Wave Compression of Condensed Matter

V.E. Fortov, L.V. Altshuler, R.F. Trunin, and A.I. Funtikov: High-Pressure Shock Compression of Solids VII

B.E. Gelfand, M.V. Silnikov, S.P. Medvedev, and S.V. Khomik: Thermo-Gas Dynamics of Hydrogen Combustion and Explosion

D. Grady: Fragmentation of Rings and Shells

Y. Horie, L. Davison, and N.N. Thadhani (Eds.): High-Pressure Shock Compression of Solids VI

J. N. Johnson and R. Cherét (Eds.): Classic Papers in Shock Compression Science

V.K. Kedrinskii: Hydrodynamics of Explosion

C.E. Needham: Blast Waves

V.F. Nesterenko: Dynamics of Heterogeneous Materials

S.M. Peiris and G.J. Piermarini (Eds.): Static Compression of Energetic Materials

M. Sućeska: Test Methods of Explosives

M.V. Zhernokletov and B.L. Glushak (Eds.): Material Properties under Intensive Dynamic Loading

J.A. Zukas and W.P. Walters (Eds.): Explosive Effects and Applications

Isabelle Sochet
Editor

Blast Effects

Physical Properties of Shock Waves

Springer

Editor
Isabelle Sochet
INSA Centre Val de Loire – PRISME Laboratory
Bourges Cedex, France

ISSN 2197-9529 ISSN 2197-9537 (electronic)
Shock Wave and High Pressure Phenomena
ISBN 978-3-319-70829-4 ISBN 978-3-319-70831-7 (eBook)
https://doi.org/10.1007/978-3-319-70831-7

Library of Congress Control Number: 2017962566

Printed on acid-free paper

This Springer imprint is published by Springer Nature
The registered company is Springer International Publishing AG
The registered company address is: Gewerbestrasse 11, 6330 Cham, Switzerland

Preface

Explosions are associated with accidental releases of energy that produce large quantities of expanding gases. Indeed, most of the past incidents involving the explosion of gas clouds suggest that a fuel leak is a potential hazard. If a quantity of fuel is accidentally released into the atmosphere and mixed with air, then a cloud of flammable gas may result. If the cloud meets an ignition source, it may develop into an explosion. However, the gas may have been initially stored as a compressed gas that was affected by a loss of containment. In all cases, a rapid expansion of gas leads to a blast or pressure wave that may have important consequences on the environment. The blast is generated by the detonation of the gas mixture unlike the pressure wave due to the deflagration of the reactive mixture. The mechanical energy of the chemical explosion creates a blast wave that moves quickly in the surrounding air.

However, we do need to focus on accidental explosion. Past and recent events naturally lead to investigate the damages caused by home-made bombs of terrorist attacks.

The shape of the blast wave depends on the type of explosion. Before the arrival of the blast wave, the pressure in the system is at normal ambient pressure. For an ideal blast wave, the pressure increases instantaneously to a maximum overpressure and then slowly decreases to negative values. Then it reaches a minimum and returns to ambient pressure. This type of blast wave, which is called a shock wave, is caused by a detonation. In addition, the blast wave has a positive phase followed by a negative phase, or a "suction phase." Although damage is most often associated with the peak of pressure, the duration and impulse of blast waves are also important parameters.

In order to offer an understanding of blast wave phenomenology, Professor John Dewey resumed in the establishment of Rankine-Hugoniot equations and the Friedlander equation, used to describe the pressure-time history of a blast wave. The primary objective of this book is to group experimental data on blast wave. A particular point is focused on the arrival time measurement and initiation of detonation by exploding wire. Gaseous and high explosive detonations are considered separately. The choice of experiments is based on the comparison of used scale sizes, from small to large scale. Each characteristic parameter of blast wave is analysed (arrival time, overpressure, positive impulse and positive duration) and expressed versus scaled distance in terms of energy and mass. A compilation of polynomial laws is given in the appendix that may be useful for engineers and researchers. A point on TNT equivalency is proposed. This study provided evidence the difficulty to determine the energy of detonation and highlighted the sensitivity of TNT equivalency to overpressure, arrival time, impulse and scaled distance. Finally, an illustration of small-scale experiments is given by Professor Gabi Ben-Dor.

Bourges Cedex, France
September 2017

Isabelle Sochet

Acknowledgements

This book is the result of the contribution of a number of PhD students and my colleagues in explosion area.

I gratefully acknowledge Professor John Dewey for the productive discussions during the MABS Symposium,[1] exchanges by mails and contributions to this book.

I gratefully acknowledge Professor Gabi Ben-Dor for allowing me to write this book, for the interesting discussions during the MABS Symposium (see footnote 1), and for his advice and contribution to this book.

I also want to offer my thanks to the Director of PRISME Laboratory, Professor Azeddine Kourta; the Director of INSA Centre Val de Loire, Professor Jean Marie Castelain; and the Research Director, Professor Frederic Kratz, who have facilitated my research activities.

I would like to acknowledge my husband Jean-François Sochet and our children Guillaume, Camille and Julien for their patience throughout the writing of this book.

[1]MABS Symposium: International Symposium on Military Aspects of Blast and Shock.

Contents

1 **Introduction** 1
Isabelle Sochet

2 **The Rankine–Hugoniot Equations: Their Extensions and Inversions Related to Blast Waves** 17
John M. Dewey

3 **The Friedlander Equations** 37
John M. Dewey

4 **Shock Wave Overpressure Measurement: Comparison Between Two Piezoelectric Sensor Materials** 57
Pietro Tadini, Kevin Gault, and Isabelle Sochet

5 **Exploding Wires** 73
Xavier Rocourt and Isabelle Sochet

6 **Blast Wave Experiments of Gaseous Charges** 89
Isabelle Sochet and Yohann Maillot

7 **Blast Wave Experiments of High Explosives** 113
Isabelle Sochet

8 **TNT Equivalency** 121
Isabelle Sochet

9 **Small-Scale Blast Wave Experiments by Means of an Exploding Wire** 141
Oren Sadot, Omri Ram, Eliram Nof, Eytan Kochavi, and Gabi Ben-Dor

Appendix A Parameters Values for Fitted Curves: Gaseous Mixtures—Hemispherical Charges 171

Appendix B Parameters Values for Fitted Curves: Gaseous Mixtures—Spherical Charges 177

Appendix C Parameters Values for Fitted Curves: High Explosives 181

Further Readings 183

Index 195

Chapter 1
Introduction

Isabelle Sochet

1.1 Definition of a Blast Wave

A blast wave is defined as the wave produced in a gas by the rapid release, or explosion, of a source of compressional energy. Examples of such sources are: the rupture of a vessel containing a compressed gas; a rapid exothermal chemical reaction, such as the detonation of a high explosive or the deflagration of a gas or vapour cloud, and a nuclear fission or fusion device. Unlike a sound wave, in which the displacements of the gas molecules are small, there is a finite displacement of the gas molecules in a blast wave. A blast wave is a non-linear wave in which the sound speed is related to the amplitude. As a result, disturbances in the wave rapidly overtake the leading edge and a characteristic shock front develops.

The term "blast wave" is used to describe the shock wave that results from the detonation of an explosive charge. This name arises from the "strong wind" that accompanies the wave and is felt by a stationary observer as the wave passes. In addition, the term "shock wave" is generally used to describe a steep pressure gradient in which the wave pattern is not characterized in detail.

1.2 Sources of Blast Waves

A blast wave is formed by the rapid expansion of the material contained in the source, which may be called the driver. In the case of the rupture of a pressurized vessel, the driver is the compressed gas within the vessel. The detonation products

I. Sochet (✉)
Laboratoire PRISME, INSA Centre Val de Loire, Bourges, France
e-mail: isabelle.sochet@insa-cvl.fr

I. Sochet (ed.), *Blast Effects*, Shock Wave and High Pressure Phenomena,
https://doi.org/10.1007/978-3-319-70831-7_1

form the driver of a chemical high explosive. In the case of a nuclear explosion, the mass of the driver materials is small relative to the amount of energy released. The contact surface or interface between the driver material and the ambient gas, which is usually the atmosphere, is unstable, and soon develops into a broad contact zone in which there is considerable mixing of the driver materials and the ambient gas. In the case of detonating explosives that are less than stoichiometric, the mixing of atmospheric oxygen with the detonation products may produce after-burning, which increases the amount of energy released from the source.

Explosions are associated with energy released that produce large quantities of expanding gases. Chemical explosions are characterized by rapid combustion accompanied by burned gas expansion. Chemical energy is partially converted into expansion that means mechanical energy. Mechanical energy of the explosion creates a blast wave that moves rapidly in the surrounding air. The resulting pressure is caused by two contradictory phenomena: increasing pressure due to combustion and decreasing pressure due to gas expansion. Blast wave represents an important property of an explosion. The blast wave affects the thermodynamic, state and dynamic of ambient air (pressure, density, particle velocity).

1.3 Unconfined Explosion Types

Three types of unconfined explosions can be defined. These categories are expressed in function of the relative position of the explosive source and the structure submitted to the blast.

- Free-air burst explosion:

  An explosion, which occurs in free air, produces an initial output whose shock wave propagates away from the center of the detonation, striking the protective structure without intermediate amplification of its wave. *UFC 3-340-02 (U.S. Army Corps of Engineers, Naval Facilities Engineering Command, Air Force Civil Engineer Support Agency 2008).*

- Air burst explosion:

  An explosion which is located at a distance from and above the protective structure so that the ground reflections of the initial wave occur prior to the arrival of the blast wave at the protective structure. As used in this manual, an air burst is limited to an explosion which occurs at two to three times the height of a one or two-story building. *UFC 3-340-2 (U.S. Army Corps of Engineers, Naval Facilities Engineering Command, Air Force Civil Engineer Support Agency 2008).*

- Surface burst explosion:

  A surface burst explosion will occur when the detonation is located close to or on the ground so that the initial shock is amplified at the point of detonation due to the ground reflections. *UFC 3-340-02 (U.S. Army Corps of Engineers, Naval Facilities Engineering Command, Air Force Civil Engineer Support Agency 2008).*

1.4 Blast Wave Profile

The expanding driver materials from a centered explosive source act like a spherical piston to produce a compressional wave in the ambient gas. The initial expansion speed of the contact surface may be supersonic, and will certainly be finite as compared to the material translation in an acoustic wave. For these reasons, the compressional wave rapidly develops a shock front at its leading edge. A shock is characterized by very rapid, and often large, changes of the physical properties of the gas through which it is passing. These changes take place typically over distances of the order of ten mean free paths of the molecules in the gas. Secondary and tertiary shocks may also develop in the blast wave arising from the repeated over expansion, contraction and re-expansion of the driver gases.

Immediately behind the primary shock front, most of the physical properties of the gas in the blast wave decay almost exponentially, and in most cases fall to values less than those in the ambient gas. An important exception to this is the case of entropy. As a shock passes through a gas there is an increase of the entropy, which is related to the strength of the shock. As the blast wave expands in three dimensions, the strength of the primary shock decreases, and leaves the ambient gas in a state of radially decreasing entropy. As a result, there is no single valued functional relationship between the thermodynamic properties, pressure and density for example, of the gases passing a fixed point enveloped by a blast wave. Thus, if a gauge is used to measure the hydrostatic pressure-time history at a fixed point in a blast wave, it is not possible to use the measured values to calculate the time-histories of any of the other physical properties such as density, temperature or particle velocity. Overcoming this feature is a significant challenge in measuring the physical properties of blast waves.

1.5 Blast Wave Characteristics

The shape of the blast wave is dependent on the nature of explosion process. Before the arrival of the blast wave, the pressure in the system is at normal ambient pressure (P_0). For an ideal blast wave, the pressure increases instantaneously to a value of $P_0 + \Delta P^+$; it then slowly decreases to negative values, reaches a minimum $P_0 - \Delta P^-$ and returns to ambient pressure. This type of blast wave, which is called a shock wave, is caused by a detonation. In the case of a deflagration, the pressure increase is slower and the maximum pressure is lower. In addition, the blast wave has a positive phase followed by a negative phase, or a "suction phase." Although damage is most often associated with the peak pressure, the duration and impulse of blast waves are also important parameters. A number of experiments conducted in recent years highlight the importance of confinement and obstacles as well as the influence of high flow speeds in the blast wave for the prediction of the characteristic parameters of an explosion, including the overpressure, impulse and duration of the positive phase.

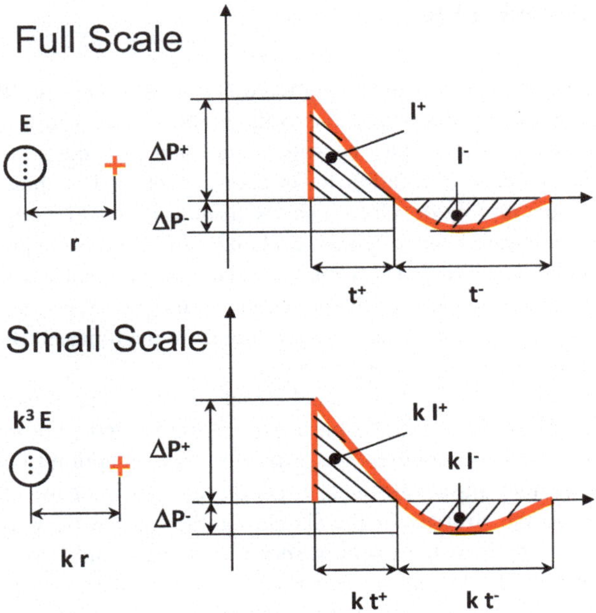

Fig. 1.1 Free field typical pressure-time variation

Thus, the change in pressure created by a detonation in free field at a fixed distance R from the center of an explosion is shown in Fig. 1.1.

In this scheme, T_a is the arrival time of the wave front, ΔP^+ is the peak hydrostatic overpressure, T^+ is the duration of the positive phase, I^+ is the positive impulse, ΔP^- is the negative depression, T^- is the duration of the negative phase and I^- is the negative impulse. The impulse I^+ of the compression phase is calculated by using the following formula:

$$I^+ = \int_{T_a}^{T_a + T^+} P(t)\, dt \tag{1.1}$$

and the impulse I^- of the rarefaction wave is given by:

$$I^- = \int_{T_a + T^+}^{T_a + T^+ + T^-} P(t)\, dt \tag{1.2}$$

1.6 Effects of Blast Waves

The increases of the hydrostatic and dynamic pressures within a blast wave are able to produce damage to structures and injury to people and animals at large distances from the source of the explosion. It is primarily for this reason that it is important to know the physical characteristics of a blast wave so that the damage and injury it may produce can be calculated. This is vital for the determination of safety distances from potential explosive sources. The destructive nature of blast waves can be put to useful purposes. Blast waves can be used to extinguish fires in situations where water cooling alone is not enough, and they provide an economical way of dismantling structures that are no longer required. Blast-type waves occur naturally, from violent volcanic eruptions for example, and are a feature of major astronomical events. Regrettably, the main sources of blast waves are from military munitions, terrorist devices and improper storage and use of energetic materials.

1.7 Energy of a Blast Wave

The characteristics of the blast wave produced by an explosive source depend on the amount of energy released, on the form of that energy and the rate at which it is released. For this reason, blast waves from different sources have different physical characteristics.

In the case of a nuclear device, approximately 50% of the energy is released as nuclear and thermal radiation, and has little direct effect on the incident blast wave. However, the thermal radiation from a nuclear device detonated on, or close to the ground, will heat the ground surface before the arrival of the blast wave. The heating of the ground can cause two phenomena that may affect the blast wave as it expands over the ground surface. The ground heated by the thermal radiation then heats a thin layer of air immediately above it. The blast wave travels more rapidly in this layer of hot air and produces a phenomenon known as a thermal precursor. In addition, the thermal radiation may be sufficiently intense to burn flammable materials, and to vapourize or violently shatter rocks and sand. This phenomenon is sometimes called "popcorning", which introduces dense material into the air ahead of the blast. The material will be transported by the flow within the blast wave and cause a significant increase of the dynamic pressure.

Another characteristic of the blast wave from a nuclear explosion is its initial very high intensity. This produces large changes of entropy in the gas through which the wave is passing. As a result of these changes, relatively less energy is available in the blast wave as it expands. In contrast, the blast wave produced by a low-grade explosion, such as the deflagration of a gas cloud, forms a low intensity wave close to the source. As the air is enveloped by such a wave, the entropy changes are small and relatively more energy is available in the wave at larger distances. For this reason, at large distances from the source, a low-grade explosive may produce a more energetic blast wave than a high-grade explosive with the same energy release.

1.8 TNT Equivalence

Most high explosives, such as TNT, RDX, HMX, C4, pentolite, nitropenta, nitroglycerin (dynamite), lead azide, silver azide and ammonium-nitrate-fuel-oil (ANFO), produce blast waves with similar, but not identical properties. As a result of the similarity, the properties of the blast waves from such explosives are often quoted in terms of the equivalent energy yield of TNT. TNT is used as the reference standard because more information is available about the properties of the blast waves produced by TNT than for any other explosive. This includes information about the properties of blast waves from so-called free-air, height-of-burst and surface-burst explosions. A free-air explosion is one produced by an energy release such that the resulting spherical blast wave does not interact with the ground or any other interface. A height-of-burst explosion is an energy release such that the incident spherical blast wave reflects from the ground producing regions of regular and of Mach reflection. Extensive information is also available about the properties of the blast waves produced by the detonation of surface-burst hemispherical charges of TNT.

It can be argued that TNT is not the ideal explosive to use as the standard. Uncased TNT cannot be detonated reliably in amounts less than about 4 kg, and thus is difficult to use in laboratory scale experiments. A more significant problem arises because TNT is chemically deficient in oxygen. As a result, TNT produces a phenomenon known as after-burning, in which the detonation products interact with atmospheric oxygen causing a lengthy release of energy, additional to that produced by the original detonation. This release of energy takes place over a time span of seconds, as compared to the detonation process that is completed in microseconds. The delayed energy release from after burning has little effect on the primary shock front or the leading part of the blast wave. However, it enhances the outward flow at later times and extends the period during which the physical properties in the blast wave are above their ambient values, particularly in the case of dynamic pressure. There are several other explosives that can be detonated, uncased, in small quantities, and which do not demonstrate after burning, but for none of these is there an extensive database of blast wave properties similar to that available for TNT.

1.9 Rate of Energy Release

Theoretically, different explosive sources can be compared on the basis of their total energy release. However, the proportion of the released energy that is contained in the resulting blast wave cannot be determined without extensive measurement. The rate of energy release also significantly affects the blast properties. For example, the energy in a vessel containing a compressed gas can be calculated from its thermodynamic properties. If the vessel ruptures, part of that energy will be used in the fracturing process and to accelerate the fragments of the container. The released gas will expand to produce a blast wave, which may further accelerate the casing fragments. As the blast wave expands in three dimensions it decelerates, but the fragments, due to their greater inertia, will decelerate less rapidly, and may penetrate and temporarily move ahead of the primary shock front. Throughout this process there is a continuous interchange of energy between the fragments and the gas contained within the blast wave.

A similar process occurs with the detonation of a high explosive in a casing, such as a munition. In this situation there is a further variability of the amount of released energy because many high explosives detonate more efficiently when cased. Because of the damage caused by the fragments from a ruptured pressure vessel or an exploding munition, the properties of the blast wave close to such an energy release are difficult to measure. At distances beyond the range of the fragments, the blast wave properties appear not to be significantly different from those from an uncased energy source of the same size.

The rate of energy release is much slower from explosive sources that rely on the use of oxygen in the atmosphere for a major part in of their energy production. Examples of such sources are the deflagration of a cloud of flammable gas, a BLEVE and specially designed thermo-baric sources. BLEVE is an acronym for Boiling Liquid Expanding Vapour Explosion. An example of a BLEVE is a propane tanker that has developed a leak. If the leaking liquid is ignited, the tanker becomes enveloped in flame and the liquid in the tanker begins to boil. The pressure caused by the expanding vapour may be enough to rupture the tanker. This preliminary explosion violently releases the propane vapour and the remaining liquid propane in the form of a droplet cloud into the atmosphere. The vapour and droplet cloud will deflagrate and detonate to produce a blast wave. The resulting blast wave will be less intense close to the source and more intense further from the source than a blast wave produced by the same energy release from a detonating high explosive.

A thermo-baric explosive is one that is designed to function in a manner similar to a BLEVE. The detonation of a primary explosive disperses a secondary explosive into the atmosphere. After mixing with atmospheric oxygen, the secondary explosive is ignited to produce a second explosion that creates a blast wave of relatively greater intensity at further distances from the source. In addition to redistributing the energy in the blast wave, a thermo-baric device may weigh less than a conventional high explosive with an equivalent energy release since it does not contain all of the oxygen needed for complete detonation.

1.10 Scaling

The distance from the source at which a specific property of a blast wave occurs, scales in proportion to the linear dimension of the source, which in turn is related to the cube root of the energy released, or of the mass of a chemical charge. For example, if a peak hydrostatic overpressure of one atmosphere occurs at a distance R from a chemical charge of mass W, the same overpressure will occur at a distance of two times from a charge of mass $8W$, two being the cube root of eight. The blast properties are also dependent on the pressure of the ambient gas. Again, a cube root scaling law applies. The time of arrival of the primary shock and the duration of it within the blast wave scale depend on the speed of sound in the ambient of gas and consequently are related to the square root of the absolute ambient temperature.

The blast wave scaling laws have been shown to apply, for most explosive sources, over many orders of magnitude of charge size. As a result, measurements of blast wave properties made using a source of a particular size can be used with confidence to determine the properties resulting from much larger or smaller energy releases from the same type of explosive. There may, however, be some limitations in the cases of some explosives. For example, some explosives do not detonate efficiently in small quantities, unless encased. TNT is such an example. An explosive such as TNT requires a relatively large booster charge of a more sensitive explosive in order for detonation to be initiated. When a small charge is used, less than about 3 kg of cast TNT for example, the force of the booster may fracture the main explosive before it is fully detonated. Fragments of the explosive will be dispersed, and some may deflagrate rather than detonate.

Another explosive for which the scaling laws must be used with care is ammonium nitrate—fuel oil (ANFO). ANFO is usually made by mixing ammonium nitrate, in the form of small, extruded particles called prill, with about 5% by mass of fuel oil. The resulting mixture is relatively stable, and can be handled easily in large quantities. ANFO cannot be detonated in quantities less than several hundred kilograms, unless cased. Attempts to use a booster charge to detonate small quantities of ANFO will result in the material being disbursed undetonated where it becomes an active fertilizer. When large quantities of uncased or lightly cased ANFO are used, the outer layer of material forms a sufficient casing for detonation to be initiated. As the quantity of ANFO is increased, so the loading density increases and detonation becomes more efficient.

1.11 Hopkinson-Cranz Laws

Hopkinson (1915) and independently Cranz (1926) have formulated the self-similar law, which is today the most common of blast scaling law.

An explosive source of mass W (or energy E) generates a blast wave with hydrostatic overpressure ΔP^+, impulse I and duration T on a point of interest located at a distance R from the center of explosion. The Hopkinson-Cranz scaling law states that at a distance $k\,R$, the same overpressure ΔP^+, the impulse $k\,I$ and duration $k\,T$ are produced by an explosive of mass $k^3\,W$ (or energy $k^3\,E$). The application of this scaling law is illustrated in Fig. 1.2. The pressure, densities, temperatures, and velocities remain unchanged at similar times in Hopkinson-Cranz scaling law.

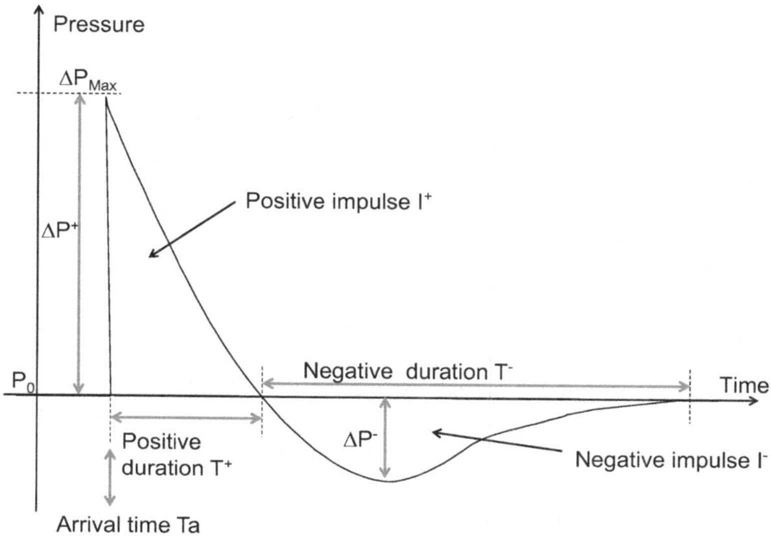

Fig. 1.2 Illustration of Hopkinson-Cranz law

Sachs (1944) has modified the Hopkinson-Cranz to take account the effect of ambient atmospheric conditions by using dimensionless parameter. Thus, the ambient atmospheric pressure P_0, ambient sound velocity a_0 are introduced in expressions of pressure P, impulse I and distance R as follows:

$$\overline{P} = \frac{P}{P_0} \quad \overline{I} = \frac{I\,a_0}{E^{1/3}\,P_0^{2/3}} \quad \overline{R} = \frac{R\,P_0^{1/3}}{E^{1/3}} \tag{1.3}$$

Dewey and Sperrazza (1950) have given the primary experimental proof of Sach's law.

1.12 Blast Interaction with Structures

The large increases of hydrostatic and dynamic pressure in a blast wave are the physical characteristics that produce most of the damage to structures and injury to people caused by blast waves. The hydrostatic pressure is defined as the pressure at a point in a gas caused by the random thermal motion of the gas molecules. It is the pressure that would be measured by a gauge moving with the flow in the gas. The dynamic pressure is a thermodynamic property of the gas defined as one half the gas density times the square of the particle speed. In an incompressible fluid the dynamic pressure is to that which would be produced if the moving fluid were brought to rest at a stagnation point. In a compressible fluid work is done to compress the gas in addition to bringing it to rest, and the stagnation pressure is not equal to the dynamic pressure.

When the shock front at the leading edge of a blast wave strikes a surface perpendicular to the direction of the flow it is reflected and a high pressure is exerted on the surface, called the reflected pressure. As the shock diffracts around the structure a rarefaction wave is produced at the edge of the structure and moves in to relieve the reflected pressure. The pressure subsequently applied to the surface is the hydrostatic pressure plus the compressible-gas stagnation pressure, known as the total pressure.

The damage produced to a structure by a blast wave depends on the structure's orientation, and is usually related to a complex combination of the hydrostatic and dynamic pressure forces. Damage to only the simplest structures can be related directly to either the hydrostatic pressure or the dynamic pressure. For example, a flat plate in a surface that is side-on to the flow will sense only the hydrostatic pressure. In contrast, a solid cylindrical rod that is oriented perpendicular to the flow may be affected only by the dynamic pressure forces.

The response of a structure exposed to a blast wave will depend also on the time-histories of the hydrostatic and dynamic pressures. A small or brittle structure, such as a glass window, may respond very rapidly to the blast loading, so that the peak pressure immediately behind the shock will be the most important criterion in determining the likely damage to the structure. In contrast, a large and massive structure will respond much more slowly, and may only be damaged if the blast loading is maintained for a long period. In such a case, the time integrals, or impulses, of the pressure forces are important damage criteria. The impulses usually quoted are the integrals of the hydrostatic and dynamic pressures from the times of arrival of the shock front to the times at which the pressures first returned to their ambient values. A damage criterion curve can be drawn for a specific type of structural damage by plotting the peak pressure and the pressure impulse at which damage just occurs for different sizes of explosive source. Such a curve typically has the form of a rectangular hyperbola in the pressure-impulse plane. The pressure and impulse at a specific distance from an explosion can be plotted on such a graph. If the point lies above the damage criterion curve, damage is likely to occur.

1.13 Effects of Height-of-Burst (HOB)

The physical properties along the ground surface from an explosive source on or close to the ground depend strongly on the height of burst. When the source is on, or very close to the ground, a significant amount of the energy may be used in the creation of a crater and of seismic waves in the ground. For this reason, the properties of the blast wave produced by a hemispherical charge on the ground surface will not be identical to those produced by a spherical mass of twice the size in the free air. As the height of burst is increased, the blast effects at ground zero (GZ), the point on the ground immediately beneath the charge, will decrease, but the effects at greater distances may increase. Therefore, there is an optimum height of burst for a given charge size to produce specified blast effects at the greatest possible distance from GZ. It was for this reason that the nuclear explosions over Hiroshima and Nagasaki were airburst explosions.

A centered explosion in free air produces a spherical blast wave. The primary spherical shock impacts the ground at GZ. As the shock continues to expand it reflects from the ground as a regular reflection in which the reflection point is on the ground surface. Because of the reflection, the hydrostatic pressure forces behind the reflected shock are greater than those behind the incident shock, but the dynamic pressure forces are less. The dynamic pressure at GZ, a stagnation point, is zero. As the reflection point moves outwards along the ground surface, a point is reached at which a regular reflection is no longer physically possible. There is then a transition to a Mach reflection in which the conjunction of the primary and reflected shocks moves above the ground and is connected to the reflecting surface by a third shock known as the Mach stem. This phenomenon is named after Ernst Mach (Mach and Sommer 1877) who was the first to describe this type of reflection.

Theory predicts a large increase of the hydrostatic pressure in the region of transition from regular to Mach reflection. This is followed by a region of enhanced dynamic pressure. The increase of hydrostatic pressure has been difficult to measure in experiments, but the enhanced dynamic pressure has been clearly demonstrated. These effects are normally presented in the form of height-of-burst curves in which the isobars for hydrostatic and dynamic pressure are plotted in the height-of-burst—ground-range plane. Because of the uncertainty which exists concerning the physical properties of blast waves in the region of transition from regular to Mach reflection, the exact form of the height-of-burst curves for explosions above the ground surface remains unresolved.

1.14 Measurement of Blast Waves

The physical properties of blast waves are normally measured using three different types of technique: electronic transducersso-called passive gauges, and high-speed photography. An electronic transducer consists of an element, such as a piezoelectric crystal, that senses a physical property in a blast wave and expresses its magnitude as an electrical output that can be amplified and recorded. The transducer is calibrated so that the electrical output can be related directly to the magnitude of the sensed physical property.

Passive gauges are inert devices that respond to the physical properties of a blast wave by deforming in some way. The final deformation can then be related to properties of the blast wave. A good example is a cantilever gauge in which a cylindrical rod exposed to the blast is either broken or permanently bent by the blast loading. Knowing the physical dimensions and the material of the cantilever, an estimate of either the peak or the impulse of the dynamic pressure can be made. Although static gauges do not provide as accurate measurements as electronic transducers, they do have some advantages. They are usually cheap, and therefore can be used in large numbers, and they can be left unattended for long periods. As a result they are useful for monitoring sites at which accidental explosions are likely to occur, and those explosions for which the exact time of the energy release is not known.

High-speed photographic techniques are probably the most powerful means of measuring the physical properties of blast waves. The primary, and sometimes the secondary shock, of a blast wave produces an optical distortion of the background when viewed from a distance. The distortion is caused by the large gradient of refractive index of the air at a shock front that results from the increased density behind the shock. High-speed photography of the distortion can be used to make accurate measurements of the radius of the shock, from which the shock speed can be calculated over a wide range of distances. The shock speed provides its Mach number, which can be used in the Rankine–Hugoniot relationships, discussed below, to calculate all of the physical properties immediately behind the shock. The analysis of the distorted images of the background caused by the change of refractive index of the ambient gas is known as refractive image analysis (RIA).

Smoke tracers introduced into the air shortly before the arrival of the blast wave can be photographed to provide the time-resolved particle trajectories within a blast wave. The particle trajectories can be used to calculate all of the other physical properties in the blast wave as functions of time and distance. Because of the state of radially decreasing entropy of the air in a blast wave, a knowledge of the time resolved particle trajectories is the only single measurement which permits a complete reconstruction of the physical properties in the wave. This reconstruction is possible because the particle trajectories in a blast wave, between the primaryand secondary shocks, made visible by flow tracers, are loci of constant entropy. As a result, the normal thermodynamic relationships can be applied to the measured physical properties along the loci. An analysis of the measured particle trajectories in a blast wave is known as particle trajectory analysis (PTA).

The impulse can be calculated by integrating the signal pressure limited to the positive or negative phase or considering the total time by using Eqs. (1.1) and (1.2). The calculation can be difficult if the signal presents some noises. For that, the pendulum devices are often used to evaluate the kinetic energy and hence the impulse transferred resulting of charge detonation (Enstock and Smith 2007). It consists of a plate which is attached to a pendulum arm. The plate submitted to the shock is deflected. The energy of the wave is transformed first into kinetic energy of the pendulum, and subsequently into potential energy (Saska et al. 2011). While Kisters et al. (2016) present a new gauge for near-field blast impulse determination: the autonomous momentum spheres (AMoS). It consists of an autonomous data recorder, an acceleration sensor and a battery cast inside a hollow sphere. From the acceleration signal of the sensors the gauge velocity–time history can be evaluated. The peak velocity of the gauge is a measure for the total impulse of the blast field. The efficiency of the autonomous blast gauges for total impulse measurement was demonstrated by tests and simulations.

1.15 Rankine–Hugoniot Relationships

Rankine and Hugoniot independently developed the Rankine–Hugoniot relationships in the later part of the nineteenth-century. The conservation of mass, momentum and energy equations were solved across a shock discontinuity to provide relationships between pairs of physical properties, such as pressure and density, across the shock. The relationships can be transformed with a Galilean translation for the case of a shock wave moving in to an ambient gas at rest. Equations can then be developed which relate each of the physical properties behind the shock to the Mach number of the shock. The Rankine–Hugoniot equations, expressed in terms of the shock Mach number, are probably the most useful tool for the measurement of blast wave properties. In addition to being used directly for the measurement of blast properties, they are invaluable for the dynamic calibration of electronic transducers. Experimental testing has proved the reliability of the Rankine-Hugoniot relationships.

The Rankine-Hugoniot equations require knowledge of the ratio of specific heats of the gas through which the shock is passing. For strong shocks, degrees of freedom of the gas molecules, in addition to their thermal motion, are excited, and this requires a modification of the ratio of specific heats in order for the Rankine-Hugoniot equations to be valid. Air can be considered to behave as an ideal gas with a ratio of specific heats equal to 1.4 for shocks with a Mach number less than about three, which corresponds to a peak hydrostatic overpressure of about ten atmospheres.

1.16 Analytical Solutions

It has not been possible to derive an analytical solution of the blast wave problem, even though the point source solution may be considered as one of the most elementary problems in fluid mechanics, viz. the instantaneous release of energy at a point a uniform medium. The first attempt was that of Sir G.I. Taylor (1946), who was able to make a solution only by assuming that the pressure in the atmosphere was infinitely small relative to that behind the primary shock. This provided good results for very intense blast waves close to the source, and enabled Taylor to make accurate predictions of the intense blast waves from the early nuclear explosions. Bethe et al. (1947) and Sakurai (1965) also attempted solutions of the point source problem.

Analytical solutions include empirical methods and semi-empirical methods. The empirical methods are essentially correlations with experimental data. Semi-empirical methods are based on simplified models of physical phenomena. They attempt to model the physical processes in a simplified way.

Characteristics of different explosives and properties of blast waves are collected in US Army Technical Manuals TM 5-1300 (U.S. Department of the Army, the Navy and the Air Force 1990) or UFC-3-340-2 (U.S. Army Corps of Engineers, Naval Facilities Engineering Command, Air Force Civil Engineer Support Agency 2008) on abacus forms.

1.17 Numerical Simulations

The numerical simulations methods (or Computational Fluid Dynamics CFD) are based on mathematical equations that describe the laws of physics governing a problem. The conservation of mass, momentum, energy, behaviour of materials, chemical reactions, etc. are written on differential or integral equations form.

The advent of high-speed computers in the 1950s made it possible to calculate the physical properties of blast waves using numerical methods, in which the ambient atmosphere is divided into a large number of cells. The explosive source becomes, in effect, a spherical piston, which impacts the cells. At a series of small time steps the influence of each cell on its neighbours is calculated. This process is iterated until an equilibrium state is found. The calculation is repeated at further time intervals until a complete mapping of the physical properties of the blast wave is obtained. Some of the first successful numerical simulations of blast waves were carried out by Brode (1955) who calculated the physical properties of blast waves from a point source, a TNT charge and a bursting pressurized glass sphere.

As the speed and capacity of computers has improved, numerous programmes have been written which accurately calculate the physical properties of blast waves from a variety of sources. Some of these, the SHARK code for example (Needham 2010) accurately calculate the complex physical and chemical phenomena that occur in and close to the explosive source. This code is based on the Low Altitude Multiple Burst Model (LAMB) procedure using the concept of image bursts. Simpler programmes assume a constant value for the ratio of specific heats of the ambient gas, but accurately predict blast wave properties for ambient peak hydrostatic pressures below about ten atmospheres. AWAF has been used to calculate a number of blast interaction configurations that form the basis of some of the figures used in this text. Numerical simulation techniques are now on the most powerful tool to predict the physical properties of blast waves from a wide variety of sources and interacting with complex structures. Experience has shown that these methods are best used in coordinated conjunction with physical measurements where now, the physical measurements are best employed to validate the numerical calculation rather than to fully characterize the blast field on their own.

Acknowledgement I would like to thank John Dewey for the discussions and contribution of this chapter.

References

Bethe, H. A., Fuchs, K., Hirschfelder, H. O., Magee, J. L., Peierls, R. E., & vonNeumann, J. (1947, August). Blast wave, LASL 2000, Los Alamos Scientific Laboratory (distributed March 27, 1958).

Brode, H. L. (1955). Numerical solutions of spherical blast waves. *Journal of Applied Physics, 26*, 766.

Cranz, C. (1926). *Lehrbuch der Ballistik*. Berlin: Springer.

Dewey, J., & Sperrazza, J. (1950). The effect of atmospheric pressure and temperature on air shock. BRL Report 721, Aberdeen Proving Ground, MD.

Enstock, L. K., & Smith, P. D. (2007). Measurement of impulse from the close-in explosion of doped charges using a pendulum. *International Journal of Impact Engineering 34*, 487–494.

Hopkinson, B. (1915). British Ordnance Board Minutes 13565.

Kisters, T., Kuder, J., & Nau, S. (2016). Autonomous gauge for blast impulse determination close to explosive charges. *Shock Waves, 26*, 117–127.

Mach, E., & Sommer, J. (1877). Uber die Fortpflanzunggeshwindigkeit von explosionsschall-wellen, Akademie der Wissenschaften, Sitzangberichte der Wiener, 74.

Needham, C. E. (2010). *Blast waves*. Berlin: Springer.

Sachs, R. G. (1944). The dependence of blast and ambient pressure and temperature. BRL Report 466.

Sakurai, A. (1965). Blast wave theory. In M. Holt (Ed.), *Basic developments in fluid mechanics* (Vol. 1, pp. 309–375). New York, NY: Academic.

Saska, P., Krzysta, E., & Mezyk, A. (2011). An analysis of an explosive shock wave impact on to military vehicles of contemporary warfare. *Journal of KONES Powertrain and Transport, 18*, 1.

Taylor, G. I. (1946). The air wave surrounding an expanding sphere. *Proceedings of the Royal Society A, 186*, 273–292.

U.S. Department of the Army, the Navy and the Air Force. (1990). Design of structures to resist the effects of accidental explosions. Technical manual TM5-1300, Washington DC.

U.S. Army Corps of Engineers, Naval Facilities Engineering Command, Air Force Civil Engineer Support Agency. (2008). Technical manuals, Unified Facilities Criteria (UFC), Structures to resist the effects of accidental explosions, UFC 3-340-02.

Chapter 2
The Rankine–Hugoniot Equations: Their Extensions and Inversions Related to Blast Waves

John M. Dewey

Symbols and Definitions

Subscripts	
s	Value of a physical property immediately behind a shock, e.g., P_s
0	Value of a physical property in the ambient gas, e.g., P_0, the ambient pressure
1	Value of a physical property in the region behind the primary shock and before the arrival of the reflected or secondary shocks, e.g., P_1
2	Value of a physical property in the region behind a reflected shock, e.g., P_2
3	Value of a physical property in the region behind a Mach stem shock, e.g., P_3
Symbols and definitions	
a	Sound speed; the speed of a weak compressional wave in a gas
a_s	Sound speed immediately behind a shock
a_0	Sound speed in an ambient gas
γ	Ratio of specific heats of a gas, i.e., $\gamma = \frac{c_P}{c_T}$; for air at NTP $\gamma = 1.4$
I_{OP+}	Impulse, or time integral, of the hydrostatic overpressure for the duration of the positive phase
P	Hydrostatic pressure, i.e., the pressure measured by a transducer that is moving with the flow, or the pressure measured by a transducer that is side-on to the flow such that it senses no normal component of the flow. It is the pressure caused by the vibrational motion of the gas molecules, only. (See also Sect. 2.3.1)
P_D	Dynamic pressure, defined as ½ ρu^2. It should be noted that in the case of a compressional flow, such as in a blast wave, the dynamic pressure is not equal to the stagnation or total pressure that is sensed by a surface perpendicular to the flow. (See also Sect. 2.3.4)

J.M. Dewey (✉)
Department of Physics and Astronomy, University of Victoria, Victoria, BC, Canada
e-mail: jdewey@uvic.ca

I. Sochet (ed.), *Blast Effects*, Shock Wave and High Pressure Phenomena,
https://doi.org/10.1007/978-3-319-70831-7_2

P_T	Total or stagnation pressure is the pressure at a point on a surface at which the gas flow has been brought to rest isentropically. (See also Sect. 2.3.5)
P_R	Normal reflected pressure is the pressure on a plane surface face-on to the shock immediately after the shock has been reflected, and before the pressure has been relieved by the arrival of the rarefaction wave initiated when the shock diffracts around the edge of the structure. (See also Sect. 2.3.6)
ρ	Density of a gas. (See also Sect. 2.3.2)
u	Gas particle velocity or flow speed. (See also Sect. 2.3.3)
r_p	The distance of air particle p from the center of the explosive source
S	Entropy of a gas, defined in terms of its change $dS \equiv dQ/T$, which results when a quantity of heat, dQ, is added to a gas at temperature T. (See also Sect. 2.3.9)
t	Time, usually measured from the moment of detonation at the center of a charge, or from the 'first light' when the detonation wave reaches the outside of the charge.
t_0	The time of arrival of the primary shock.
t^+	The positive duration of a physical property of a blast wave from the time of arrival of the primary shock to the time when the property first returns to its ambient value. (See also Sect. 3.7)

2.1 Introduction

The Rankine–Hugoniot equations were developed in their original forms independently by Rankine (1870a, b) and Hugoniot (1887, 1889). The equations describe the relationships between the physical properties in the two possible states of a moving compressible gas for which mass, momentum and energy are conserved. The contact surface between these two states is a shock front. In their original forms, the equations described the relationships for a supersonic gas passing through a stationary shock into a subsonic state. For blast wave applications, the equations are transformed to describe the physical properties of the gas behind a shock moving into a stationary ambient atmosphere, in terms of the Mach number of the shock.

The Rankine–Hugoniot relationships have been in use for more than a century, and subject, therefore, to intense evaluation. All such evaluations have demonstrated the absolute validity of the relationships, assuming the correct value of the ratio of specific heats has been used. The relationships have the same order of reliability as the classical equations of thermodynamics. As a result, the most reliable method of calibrating gauges and transducers used to measure the physical properties of blast waves is to measure the shock speed using two simple time-of-arrival detectors; calculate the shock Mach number, and then use this in the appropriate Rankine–Hugoniot relationship to determine the change of the physical property across the shock. This value can then be related to the output of the transducer.

For the study of blast waves, the primary shock usually is moving into the air of the ambient atmosphere, which can be considered as a perfect gas with a ratio of specific heats, $\gamma = 1.4$. When the overpressure ratio across the primary shock is greater than about ten atmospheres, corresponding to a shock Mach number of about three, additional degrees of freedom are excited in the air molecules, and the ratio of specific heats can no longer be considered constant. For shocks with a Mach number greater than 3.5, i.e., a hydrostatic overpressure of 13 atm (193 psi, 1330 kPa), an assumed $\gamma = 1.4$ leads to an error of more than 1%. For shocks with a Mach number greater than 5.5, i.e., a hydrostatic overpressure of about 34 atm (500 psi, 3458 kPa), the error is more than 5%.

The most commonly used forms of the equations relate the hydrostatic overpressure, density, and particle velocity to the shock Mach number. In this chapter, these equations are extended to give eleven different physical properties in terms of the ratio of specific heats, γ, and for $\gamma = 1.4$. In the latter case, the equations are also inverted to provide the shock Mach number, and thus the other ten properties, in terms of a specified physical property.

2.2 The Original Rankine–Hugoniot Relationships

Consider a perfect gas moving in a cylinder of constant cross-section and passing from state **1** to state **2** through a stationary discontinuity, as shown in Fig. 2.1. In state **1** the hydrostatic pressure, density, and particle speed are P_1, ρ_1, and w_1, respectively, and w_1 is supersonic. In state **2** the corresponding properties are P_2, ρ_2, and w_2, and w_2 is subsonic.

The equations representing the conservation of mass, momentum, and energy across the discontinuity may be expressed, respectively, as

$$\rho_2 w_2 = \rho_1 w_1, \tag{2.1}$$

$$P_2 + \rho_2 w_2^2 = P_1 + \rho_1 w_1^2, \text{ and} \tag{2.2}$$

$$\frac{P_2}{\rho_2} + e_2 + \frac{1}{2} w_2^2 = \frac{P_1}{\rho_1} + e_1 + \frac{1}{2} w_1^2, \tag{2.3}$$

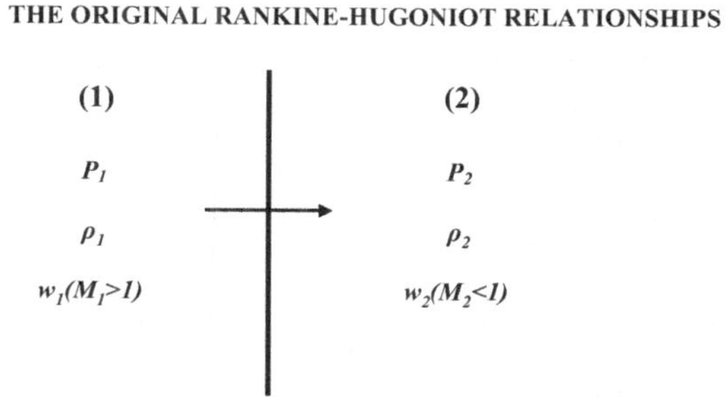

Fig. 2.1 A perfect gas moving through a stationary discontinuity from state **1** to state **2**. The relevant physical properties of the gas are: hydrostatic pressure P, density ρ, and particle speed w

where e is the internal energy of the gas, per unit volume.

The enthalpy of the gas, h, is defined as

$$h = \frac{P}{\rho} + e. \tag{2.4}$$

For a thermally and calorically perfect gas

$$P = \rho RT, \text{ and} \tag{2.5}$$

$$h = \frac{\gamma}{\gamma - 1} RT, \tag{2.6}$$

where R is the universal gas constant and T the absolute temperature.

Using (2.1)–(2.6), the following relationships can be derived,

$$\frac{P_2}{P_1} = \frac{1 - \frac{\rho_1}{\rho_2}\left(\frac{\gamma-1}{\gamma+1}\right)}{\frac{\rho_1}{\rho_2} - \frac{\gamma-1}{\gamma+1}}, \text{ and} \tag{2.7}$$

$$\frac{\rho_2}{\rho_1} = \frac{\frac{\gamma-1}{\gamma+1} + \frac{P_2}{P_1}}{\frac{\gamma-1}{\gamma+1}\frac{P_2}{P_1} + 1} = \frac{w_1}{w_2}. \tag{2.8}$$

These relationships may also be written in terms of the Mach number of the flow in region 1, M_1, where

$$M_1 \equiv \frac{w_1}{a_1}, \text{ and} \tag{2.9}$$

a_1, the speed of sound in region 1 is given by

$$a_1^2 = \gamma \frac{P_1}{\rho_1}. \tag{2.10}$$

2.3 The Rankine–Hugoniot Equations for a Blast Wave Shock

At this point it is convenient to change the nomenclature for the situation in which a shock, with speed V_S, is moving into an ambient atmosphere at rest. Suffix *0* will be used to identify the physical properties of the ambient gas, and suffix *S* to identify those immediately behind the shock, as shown in Fig. 2.2. For most blast waves, these "peak" values decay rapidly in an almost exponential fashion, as functions of both time and distance.

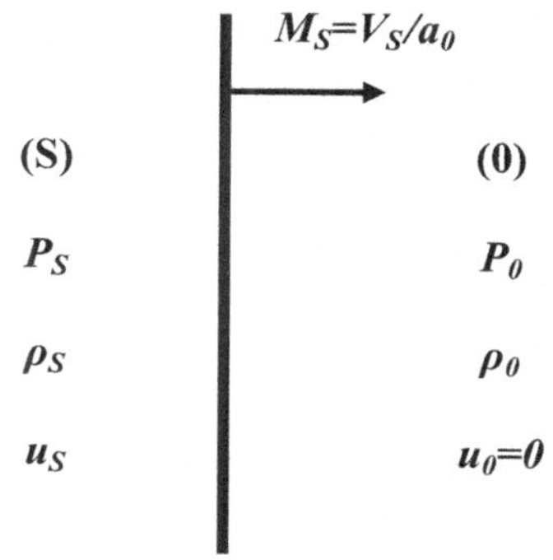

Fig. 2.2 Configuration of a shock moving with speed V_S into a stationary gas in state **0**. The gas behind the shock is in state **S**. P, ρ, and u are the hydrostatic pressure, density, and particle speed, respectively. The shock Mach number $M_S = V_S/a_0$, where a_0 is the ambient sound speed

Applying these changes to (2.7)–(2.10) gives,

$$\frac{P_S}{P_0} = \frac{2\gamma M_S^2 - (\gamma - 1)}{\gamma + 1}, \text{ and} \tag{2.11}$$

$$\frac{\rho_S}{\rho_0} = \frac{(\gamma + 1) M_S^2}{(\gamma - 1) M_S^2 + 2}. \tag{2.12}$$

2.3.1 *Hydrostatic Pressure*

The hydrostatic pressure is the pressure measured by a transducer moving with the flow, or the pressure measured by a transducer that is side-on to the flow such that it senses no component of the flow normal to the transducer. It is the pressure caused by the vibrational motion of the gas molecules, only. Some texts refer to this as the static pressure, but here the term hydrostatic pressure is used to emphasize the difference from the total or stagnation pressure. The hydrostatic pressure in excess of the ambient pressure is the hydrostatic overpressure.

When describing blast waves, hydrostatic pressure is usually measured in units of kiloPascals (kPa), pounds-weight per square inch (psi), or nondimensionally in atmospheres (atm). An informal tripartite (USA, UK, and Canada) agreement recommended that the standard atmosphere to be used when describing blast waves is that at NTP (*N*ormal *T*emperature and *P*ressure), viz. 15 °C (288.16 K) and 101.325 kPa (14.696 psi). This was considered to be more appropriate than STP (*S*tandard *T*emperature and *P*ressure) for which the standard temperature is 0 °C. The hydrostatic overpressure, OP_S, the pressure in excess of the ambient pressure P_0, with (2.11) gives

$$\frac{\mathrm{OP}_S}{P_0} = \frac{P_S - P_0}{P_0} = \frac{P_S}{P_0} - 1 = \frac{2\gamma M_S^2 - (\gamma - 1)}{\gamma + 1} - 1 = \frac{2\gamma}{\gamma + 1}\left(M_S^2 - 1\right). \tag{2.13}$$

For $\gamma = 1.4$, (2.11) becomes

$$\frac{P_S}{P_0} = \frac{7M_S^2 - 1}{6}, \tag{2.14}$$

and (2.13) becomes

$$\frac{OP_S}{P_0} = \frac{7}{6}\left(M_S^2 - 1\right). \tag{2.15}$$

The inverses of (2.14) and (2.15), which give the shock Mach number in terms of the pressures, are, respectively,

$$M_S = \sqrt{\frac{1}{7}\left(\frac{6P_S}{P_0} + 1\right)}, \text{ and} \tag{2.16}$$

$$M_S = \sqrt{\frac{6}{7}\frac{\mathrm{OP}_S}{P_0} + 1}. \tag{2.17}$$

2.3.2 Density

Density (ρ) is the mass per unit volume of a gas, measured in units of kilograms per cubic meter (kg m^{-3}), pounds-mass per cubic foot (lb ft^{-3}), or pounds-mass per cubic inch (lb in^{-3}). Nondimensionally, it may be measured relative to the density of air at NTP, which is 1.225 kg m^{-3} (0.076475 lb ft^{-3}, 4.4256×10^{-5} lb in^{-3}). When a gas is traversed by a shock, there is a rapid increase of the density. For an ideal gas with a ratio of specific heats of $\gamma = 1.4$, there is an upper limit of 6.0 for the ratio of the densities, ρ_S/ρ_0, across an infinitely strong shock. In practice, very strong shocks produce other changes to the gas so that the ratio of specific heats, γ, does not remain constant, and larger density changes can occur.

The density ratio across the shock is given by (2.12) in terms of the shock Mach number, as

$$\frac{\rho_S}{\rho_0} = \frac{(\gamma + 1)M_S^2}{(\gamma - 1)M_S^2 + 2}. \tag{2.18}$$

For $\gamma = 1.4$

$$\frac{\rho_S}{\rho_0} = \frac{6M_S^2}{M_S^2 + 5}, \text{ and} \tag{2.19}$$

the inverse is

$$M_S = \sqrt{\frac{5\rho_1/\rho_0}{6 - \rho_1/\rho_0}}. \tag{2.20}$$

2.3.3 Particle Velocity

Particle velocity (u) is the translational velocity of the gas within a blast wave, measured in meters per second (m s^{-1}) or feet per second (ft s^{-1}). Nondimensionally, the particle velocity may also be quoted relative to either the sound speed in the ambient gas (a_0), or the sound speed at the location of the gas particle (a). u/a_0 is not a Mach number, but a dimensionless quantity that is useful in scaling blast waves for different charge sizes and atmospheric conditions. The particle velocity in terms of the sound speed at the same position in the blast wave, u/a, is a true Mach number and is known as the local Mach number of the flow. When considering the blast interaction with a structure, it is important to know if the local Mach number is less than or greater than one. If the local Mach number is greater than one, i.e., supersonic, a bow shock forms around the structure and this further changes the properties of the gas before it can interact with the structure, as described in the section on total pressure, below.

The coordinate transformation for a shock moving into a stationary gas and the corresponding change of nomenclature, using (2.8), gives

$$\frac{\rho_S}{\rho_0} = \frac{w_1}{w_2} = \frac{V_S}{V_S - u_S} = \frac{M_S}{M_S - u_S/a_0}, \tag{2.21}$$

where a_0 is the sound speed in the ambient gas. Thus, using (2.21) and (2.18),

$$M_S - \frac{u_S}{a_0} = M_S \frac{\rho_0}{\rho_S}, \text{ and}$$

$$\begin{aligned} \frac{u_S}{a_0} = -M_S\left(\frac{\rho_0}{\rho_S} - 1\right) &= -M_S\left[\frac{(\gamma - 1)M_S^2 + 2}{(\gamma + 1)M_S^2} - 1\right] \\ &= \frac{2}{\gamma + 1}\left(\frac{M_S^2 - 1}{M_S}\right). \end{aligned} \tag{2.22}$$

For $\gamma = 1.4$, (2.22) becomes

$$\frac{u_S}{a_0} = \frac{5}{6}\left(\frac{M_S^2 - 1}{M_S}\right), \text{ and} \tag{2.23}$$

the inverse of (2.23) is

$$M_S = \frac{1}{5}\left(3\frac{u_S}{a_0} + \sqrt{9\left(\frac{u_S}{a_0}\right)^2 + 25}\right). \tag{2.24}$$

2.3.4 Temperature

Absolute temperature (T) is the temperature of a gas measured from the absolute zero, −273.16 °C, in Kelvin (K). The absolute temperature at NTP is therefore 288.16 K. The absolute temperature of the gas behind the shock, T_S, is obtained in terms of the ambient temperature, T_0, using (2.5), (2.14), and (2.18) as

$$\begin{aligned}\frac{T_S}{T_0} = \frac{P_S}{\rho_S}\frac{\rho_0}{P_0} = \frac{P_S}{P_0}\frac{\rho_0}{\rho_S} &= \frac{2\gamma M_S^2 - (\gamma-1)}{\gamma+1}.\frac{(\gamma-1)M_S^2+2}{(\gamma+1)M_S^2} \\ &= \frac{\left[2\gamma M_S^2-(\gamma-1)\right].\left[(\gamma-1)M_S^2+2\right]}{(\gamma+1)^2M_S^2}.\end{aligned} \tag{2.25}$$

For $\gamma = 1.4$,

$$\frac{T_S}{T_0} = \frac{\left(7M_S^2-1\right)\left(M_S^2+5\right)}{36M_S^2}, \tag{2.26}$$

and the inverse is

$$M_S = \sqrt{\frac{1}{14}\left[\left(36\frac{T_S}{T_0}-34\right)+\sqrt{\left(34-36\frac{T_S}{T_0}\right)^2+140}\right]}. \tag{2.27}$$

2.3.5 Sound Speed

The sound speed of a gas (a) is the speed at which a weak compression or rarefaction wave is transmitted through the gas. For one mole of an ideal gas $P/\rho = RT$, where R is the universal gas constant, and the speed of sound is given by $a = \surd(\gamma P/\rho) = \surd(\gamma RT)$. Therefore, the speed of sound is proportional to the square root of the absolute temperature. Using the above, the sound speed in an ambient atmosphere at NTP is 340.4288 m/s or 1116.878 ft/s. The sound speed immediately behind the shock, a_S, in terms of the ambient sound speed, a_0, is obtained from (2.26) as,

$$\frac{a_S}{a_0} = \sqrt{\frac{T_S}{T_0}} = \frac{1}{(\gamma+1)M_S}\sqrt{\left[2\gamma M_S^2-(\gamma-1)\right].\left[(\gamma-1)M_S^2+2\right]}. \tag{2.28}$$

For $\gamma = 1.4$, (2.28) becomes

$$\frac{a_S}{a_0} = \frac{1}{6M_S}\sqrt{\left(7M_S^2-1\right)\left(M_S^2+5\right)}, \tag{2.29}$$

and the inverse of (2.29) is

$$M_S = \sqrt{\frac{1}{14}\left\{\left[36\left(\frac{a_S}{a_0}\right)^2 - 34\right] + \sqrt{\left[34 - 36\left(\frac{a_S}{a_0}\right)^2\right]^2 + 140}\right\}}. \tag{2.30}$$

2.3.6 Local Mach Number

The Mach number of the flow immediately behind the shock, u_s/a_s, using (2.22) and (2.28), is

$$\begin{aligned}\frac{u_S}{a_S} = \frac{u_S}{a_0}\frac{a_0}{a_S} &= \frac{2}{\gamma+1}\left(\frac{M_S^2-1}{M_S}\right).\frac{(\gamma+1)\,M_S}{\sqrt{\left[2\gamma M_S^2-(\gamma-1)\right].\left[(\gamma-1)\,M_S^2+2\right]}}\\ &= \frac{2\left(M_S^2-1\right)}{\sqrt{\left[2\gamma M_S^2-(\gamma-1)\right].\left[(\gamma-1)\,M_S^2+2\right]}}.\end{aligned} \tag{2.31}$$

For $\gamma = 1.4$,

$$\frac{u_S}{a_S} = \frac{5\left(M_S^2-1\right)}{\sqrt{\left(7M_S^2-1\right)\left(M_S^2+5\right)}}, \tag{2.32}$$

and the inverse of (2.32) is

$$M_S = \sqrt{\frac{17\left(\frac{u_S}{a_S}\right)^2 + 25 + 6\sqrt{9\left(\frac{u_S}{a_S}\right)^4 + 25\left(\frac{u_S}{a_S}\right)^2}}{25 - 7\left(\frac{u_S}{a_S}\right)^2}}. \tag{2.33}$$

2.3.7 Dynamic Pressure

Dynamic pressure (P_D) is defined as one half the gas density times the square of the particle velocity, i.e., ½ $\rho\, u^2$. Dynamic pressure is a scalar property of the gas, and since gas is a compressible fluid, is not equal to the stagnation or total pressure exerted on a surface at which the gas is brought to rest, for which see below. Although the dynamic pressure does not represent the pressure exerted on any surface in the blast wave, it is a useful measure of the relative importance of the drag forces produced by the wave as compared to the hydrostatic forces. The dynamic pressure immediately behind a shock, using Eqs. (2.18) and (2.22), is given in terms of the ambient pressure, as

$$\frac{P_{\mathrm{DS}}}{P_0} = \frac{1}{2}\frac{\rho_S u_S^2}{P_0} = \frac{1}{2}\frac{\rho_S}{\rho_0}\left(\frac{u_S}{a_0}\right)^2\frac{\rho_0 a_0^2}{P_0} = \frac{1}{2}\frac{\rho_S}{\rho_0}\left(\frac{u_S}{a_0}\right)^2\frac{\rho_0}{P_0}\frac{\gamma P_0}{\rho_0} = \frac{1}{2}\gamma\frac{\rho_S}{\rho_0}\left(\frac{u_S}{a_0}\right)^2$$
$$= \frac{1}{2}\gamma\frac{(\gamma+1)M_S^2}{(\gamma-1)M_S^2+2}\cdot\left[\frac{2}{\gamma+1}\left(\frac{M_S^2-1}{M_S}\right)\right]^2$$
$$= \frac{2\gamma}{\gamma+1}\frac{\left(M_S^2-1\right)^2}{(\gamma-1)M_S^2+2}. \tag{2.34}$$

For $\gamma = 1.4$,

$$\frac{P_{\mathrm{DS}}}{P_0} = \frac{35}{12}\frac{\left(M_S^2-1\right)^2}{M_S^2+5}, \tag{2.35}$$

and the inverse is

$$M_S = \sqrt{\frac{1}{70}\cdot\left(12\frac{P_D}{P_0}+70+\sqrt{\left(12\frac{P_D}{P_0}+70\right)^2+700\left(12\frac{P_D}{P_0}-7\right)}\right)}. \tag{2.36}$$

N.B. Dynamic pressure, as defined above, is a scalar property. In the study of blast waves it is used to describe the drag forces on structures, for which the direction of the flow is important and therefore is better defined as a vector, viz. ½ $\rho\, u\, |u|$. The relationships (2.34), (2.35), and (2.36) remain valid with this revised definition.

2.3.8 *Total Overpressure*

Total overpressure (OP_T), or stagnation overpressure, is defined as the increase of pressure sensed by a transducer or a surface which is face-on to the flow, as a result of the flow being brought to rest isentropically. Work is done both to bring the gas to rest and to compress it adiabatically. The total overpressure is not the same as the reflected overpressure, for which see below, or the dynamic pressure, see above.

When the primary shock of a blast wave strikes a rigid surface face on, it will be reflected, and the reflected shock will run back into the flow, as shown in Fig. 2.3. If the flow produced by the primary shock is subsonic, i.e., $u_S/a_S < 1$, the reflected shock will continue to move back through the flow and will disperse. In this case, the face-on rigid surface will continue to be exposed to the decaying flow in region 1 behind the primary shock. If the flow produced by the primary shock is supersonic, i.e., $u_S/a_S > 1$, the reflected shock will move back into the flow until a point is reached when its speed is equal to that of the flow, i.e., $M_R = V_R/a_S = u_S/a_S$. It will then form a bow shock through which the flow must pass before being stagnated at

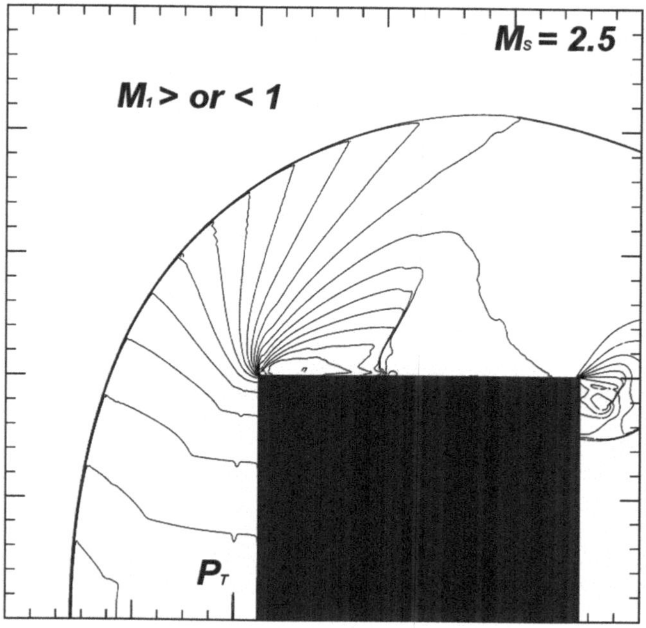

Fig. 2.3 Numerical simulation showing the shock reflected from a rigid structure. The Mach number of the incident shock, $M_S = 2.5$. The total pressure, P_T, is the stagnation pressure on the front face of the structure. The contours are isopyknics. (Courtesy A. A. van Netten)

the rigid surface. The physical properties of the gas are changed as it passes through the bow shock and these changes must be taken into consideration when calculating the total overpressure.

2.3.8.1 Subsonic Case

The overpressure at a stagnation point in a gas may be written as (Prandtl and Tietjens 1934, p. 227)

$$q = p\left(1 + \frac{\rho w^2}{2p}\frac{\gamma - 1}{\gamma}\right)^{\frac{\gamma}{\gamma-1}} - 1, \tag{2.37}$$

where p is the pressure in the free stream, ρ the density, w the flow speed, and γ the ratio of specific heats.

Writing (2.37) in terms of the flow conditions immediately behind a shock gives the total overpressure, $\mathrm{OP_{TS}}$, as

$$\mathrm{OP_{TS}} = P_S\left(1 + \frac{\rho_S u_S^2}{2P_S}\frac{\gamma - 1}{\gamma}\right)^{\frac{\gamma}{\gamma-1}} - 1. \tag{2.38}$$

Writing the physical properties as ratios of their values in the ambient atmosphere, (2.38) becomes

$$\begin{aligned} \frac{\mathrm{OP_{TS}}}{P_0} &= \frac{P_S}{P_0}\left(1+\frac{1}{2}\frac{\rho_S}{\rho_0}\frac{u_S^2}{a_0^2}\frac{P_0}{P_S}\frac{\rho_0 a_0^2}{P_0}\frac{\gamma-1}{\gamma}\right)^{\frac{\gamma}{\gamma-1}}-1 \\ &= \frac{P_S}{P_0}\left(1+\frac{1}{2}\frac{\rho_S}{\rho_0}\left[\frac{u_S}{a_0}\right]^2\frac{P_0}{P_1}[\gamma-1]\right)^{\frac{\gamma}{\gamma-1}}-1, \end{aligned} \tag{2.39}$$

since $\dfrac{\rho_0 a_0^2}{P_0}=\dfrac{\rho_0}{P_0}\dfrac{\gamma P_0}{\rho_0}=\gamma$. Writing the ratios across the shock in terms of the shock Mach number, M_S, using (2.14), (2.18), and (2.22), gives

$$\begin{aligned} \frac{\mathrm{OP_{TS}}}{P_0} &= \frac{2\gamma M_S^2-(\gamma-1)}{\gamma+1} \\ &\quad \left[1+\frac{(\gamma+1)M_S^2}{(\gamma-1)M_S^2+2}\left\{\frac{2}{\gamma+1}\left(\frac{M_S^2-1}{M_S}\right)\right\}^2\frac{(\gamma+1)(\gamma-1)}{2\gamma M_S^2-(\gamma-1)}\right]^{\frac{\gamma}{\gamma-1}}-1 \\ &= \frac{2\gamma M_S^2-(\gamma-1)}{\gamma+1}\left[\frac{M_S^2\{2(\gamma^2-1)M_S^2-(\gamma-3)(\gamma+1)\}}{\{(\gamma-1)M_S^2+2\}\{2\gamma M_S^2-(\gamma-1)\}}\right]^{\frac{\gamma}{\gamma-1}}-1. \end{aligned} \tag{2.40}$$

For $\gamma = 1.4$,

$$\frac{\mathrm{OP_{TS}}}{P_0} = \frac{7M_S^2-1}{6}\left[\frac{12M_S^2(M_S^2+2)}{(M_S^2+5)(7M_S^2-1)}\right]^{3.5}-1. \tag{2.41}$$

It has not been possible to find a simple analytical inversion of (2.41), but the following empirical relationship provides a good description of the shock Mach number in terms of the total overpressure in the subsonic case,

$$M_S = 1.0816-0.06119\frac{\mathrm{OP_{TS}}}{P_0}+1.0903\left(\frac{\mathrm{OP_{TS}}}{P_0}\right)^{1/2}-0.81528\left(\frac{\mathrm{OP_{TS}}}{P_0}\right)^{1/3}. \tag{2.42}$$

2.3.8.2 Supersonic Case

Equation (2.31) gives the local Mach number of the flow behind a shock in terms of the shock Mach number, M_S, and the ratio of specific heats, γ. Putting the flow Mach number equal to 1, and $\gamma = 1.4$ gives a value for the shock Mach number of 2.0681. For shocks stronger than this value, the flow immediately behind the shock is locally supersonic. In this case, a bow shock forms ahead of any stationary object or structure enveloped by the primary shock of the blast wave. The supersonic flow will pass through the bow shock and be made locally subsonic before being brought to rest by the object or structure.

The total pressure behind a reflected shock in terms of the hydrostatic pressure, sometimes called the Rayleigh supersonic pitot formula, is given as Eq. (6.3), in Liepmann and Roshko (1957). Using this equation gives the total overpressure, $\mathrm{OP_{TS}}$, as

$$\frac{OP_{\mathrm{TS}}}{P_0} = \frac{P_{T2}}{P_0} - 1 = \frac{P_{T2}}{P_S}\frac{P_S}{P_0} - 1 = \frac{\left(\frac{\gamma+1}{2}M_1^2\right)^{\frac{\gamma}{\gamma-1}}}{\left(\frac{2\gamma}{\gamma+1}M_1^2 - \frac{\gamma-1}{\gamma+1}\right)^{\frac{1}{\gamma-1}}}\frac{P_S}{P_0} - 1, \tag{2.43}$$

where P_{T2} is the total pressure behind the reflected shock, P_0 is the pressure of the ambient atmosphere ahead of the incident shock, and P_S and M_1 are the hydrostatic pressure and flow Mach number, respectively, behind the incident shock.

Using (2.13) and (2.31) in (2.43) gives

$$\frac{\mathrm{OP_{TS}}}{P_0} = \frac{\left[\frac{2(\gamma+1)\left(M_S^2-1\right)^2}{\left(2\gamma M_S^2-\gamma+1\right)\left((\gamma-1)M_S^2+2\right)}\right]^{\frac{\gamma}{\gamma-1}}\left[\frac{2\gamma\left(M_S^2-1\right)}{\gamma+1}+1\right]}{\left[\frac{8\gamma\left(M_S^2-1\right)^2}{(\gamma+1)\left(2\gamma M_S^2-\gamma+1\right)\left((\gamma-1)M_S^2+2\right)} - \frac{\gamma-1}{\gamma+1}\right]^{\frac{1}{\gamma-1}}} - 1. \tag{2.44}$$

For $\gamma = 1.4$,

$$\frac{\mathrm{OP_{TS}}}{P_0} = \frac{67920.1\left(M_S^2-1\right)^7}{\left(M_S^2+5\right)\left(42M_S^4-96M_S^2+45\right)^{2.5}} - 1. \tag{2.45}$$

No analytical solution of (2.45) could be found to give the shock Mach number in terms of the total pressure in the supersonic case, but the following empirical relationship, (2.46), gives a good description for shock Mach numbers from 2.06 to 3.5. For shock Mach numbers greater than 3.5, real-gas effects begin to become important, and these equations become increasingly less accurate.

$$M_S = 1.1421 + 0.3091\left(\frac{\mathrm{OP_{TS}}}{P_0}\right)^{0.5305}. \tag{2.46}$$

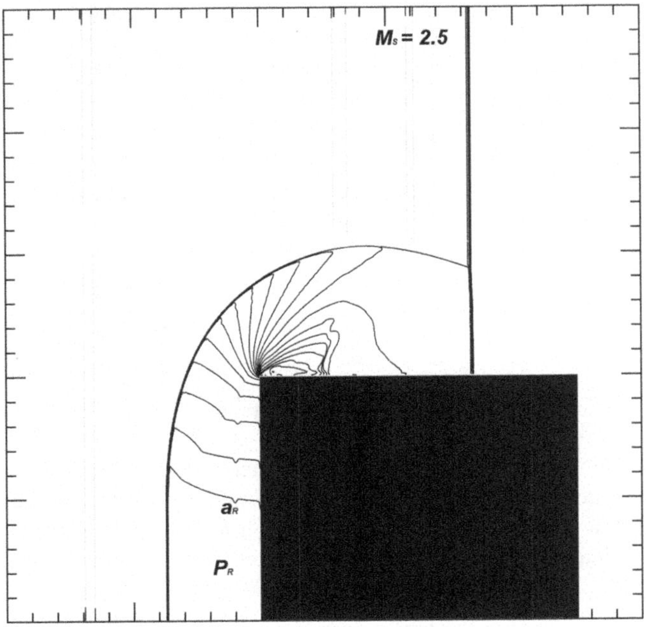

Fig. 2.4 Numerical simulation of a Mach 2.5 shock incident on a rigid structure. The gas behind the reflected shock has been brought to rest non-isentropically, and the pressure on the structure is the reflected pressure, P_R. As the reflected shock diffracts around the edge of the structure, a rarefaction wave is produced which moves at the speed of sound behind the reflected shock, a_R. The contours are isopyknics. (Courtesy A. A. van Netten)

2.3.9 Reflected Overpressure

Reflected overpressure ($\mathrm{OP_{RS}}$) is the overpressure exerted on a plane surface face-on to the shock front, immediately after the shock reflection. If the plane, reflecting surface is finite in size, the reflected overpressure will be relieved by a rarefaction wave generated as the reflected shock diffracts around the boundary of the reflecting surface (Fig. 2.4). The reflected pressure, P_{RS}, is given by

$$\frac{P_{\mathrm{RS}}}{P_0} = \left\{\frac{2\gamma M_S^2 - (\gamma - 1)}{\gamma + 1}\right\}\left\{\frac{(3\gamma - 1)\,M_S^2 - 2\,(\gamma - 1)}{(\gamma - 1)\,M_S^2 + 2}\right\}, \tag{2.47}$$

and the reflected overpressure, OP_{RS}, is

$$\frac{\mathrm{OP_{RS}}}{P_0} = \frac{P_{\mathrm{RS}}}{P_0} - 1 = \left\{\frac{2\gamma M_S^2 - (\gamma - 1)}{\gamma + 1}\right\}\left\{\frac{(3\gamma - 1)\,M_S^2 - 2\,(\gamma - 1)}{(\gamma - 1)\,M_S^2 + 2}\right\} - 1. \tag{2.48}$$

For $\gamma = 1.4$,

$$\frac{\mathrm{OP_{RS}}}{P_0} = \frac{\left(7M_S^2 - 1\right)\left(4M_S^2 - 1\right)}{3\left(M_S^2 + 5\right)} - 1, \tag{2.49}$$

and the inverse of (2.49) is

$$M_S = \frac{1}{128}\sqrt{196 + 42\left(\frac{\mathrm{OP_{RS}}}{P_0}\right) + 42\sqrt{196 + 196\left(\frac{\mathrm{OP_{RS}}}{P_0}\right) + \left(\frac{\mathrm{OP_{RS}}}{P_0}\right)^2}}. \tag{2.50}$$

2.3.10 Reflected Temperature and Sound Speed

It is necessary to know the sound speed behind a reflected shock in order to calculate the time for the rarefaction wave, produced when the reflected shock diffracts around the edges of the reflecting surface, to move across the reflecting surface. After the arrival of the rarefaction wave, the excess pressure on the surface will reduce to the total overpressure, $\mathrm{OP_{TS}}$, described above.

The absolute temperature, T_R, and sound speed, a_R, in the region behind a reflected shock are derived in terms of the incident shock Mach number, M_S, as follows, using (2.11) and (2.47)

$$\begin{aligned}\frac{T_{\mathrm{RS}}}{T_0} &= \frac{\gamma P_{\mathrm{RS}}}{\rho_{\mathrm{RS}}}\frac{\rho_0}{\gamma P_0} = \frac{P_{\mathrm{RS}}}{P_0}\frac{\rho_0}{\rho_R}\\ &= \frac{P_{\mathrm{RS}}}{P_0}\left(1 + \frac{\gamma-1}{\gamma+1}\frac{P_{\mathrm{RS}}}{P_0}\right)\Big/\left(\frac{\gamma-1}{\gamma+1} + \frac{P_{\mathrm{RS}}}{P_0}\right)\\ &= \frac{\left[2(\gamma-1)M_S^2 + (3-\gamma)\right]\left[(3\gamma-1)M_S^2 - 2(\gamma-1)\right]}{(\gamma+1)^2 M_S^2}, \text{ and}\end{aligned} \tag{2.51}$$

$$\frac{a_{\mathrm{RS}}}{a_0} = \sqrt{\frac{T_{\mathrm{RS}}}{T_0}} = \frac{\sqrt{\left[2(\gamma-1)M_S^2 + (3-\gamma)\right]\left[(3\gamma-1)M_S^2 - 2(\gamma-1)\right]}}{(\gamma+1)M_S}. \tag{2.52}$$

For $\gamma = 1.4$,

$$\frac{T_{\mathrm{RS}}}{T_0} = \frac{\left(M_S^2+2\right)\left(4M_S^2-1\right)}{9M_S^2}, \text{ and} \tag{2.53}$$

$$\frac{a_R}{a_0} = \frac{\sqrt{\left(M_S^2+2\right)\left(4M_S^2-1\right)}}{3M_S}. \tag{2.54}$$

The inverse of (2.53) is

$$M_S = \frac{1}{4}\sqrt{18\frac{T_{\mathrm{RS}}}{T_0} - 14 + 6\sqrt{9\left(\frac{T_{\mathrm{RS}}}{T_0}\right)^2 - 14\frac{T_{\mathrm{RS}}}{T_0} + 9}}, \tag{2.55}$$

and of (2.54) is

$$M_S = \frac{1}{4}\sqrt{18\left(\frac{a_{RS}}{a_0}\right)^2 - 14 + 6\sqrt{9\left(\frac{a_{RS}}{a_0}\right)^4 - 14\left(\frac{a_{RS}}{a_0}\right)^2 + 9}}. \tag{2.56}$$

2.3.11 Entropy

The entropy (S) of a gas is defined in terms of its change, dS, when a quantity of heat, dQ, is added at a temperature T, viz.

$$\mathrm{d}S \equiv \mathrm{d}Q/T. \tag{2.57}$$

Since $\mathrm{d}Q = c_v\,\mathrm{d}T + P\mathrm{d}V$, for a perfect gas, the total change of entropy across a shock is given by

$$\Delta S = \int \frac{\mathrm{d}Q}{T} = c_v \int_{T_0}^{T_1} \frac{\mathrm{d}T}{T} + R\int_{V_0}^{V_1} \frac{\mathrm{d}V}{V}, \tag{2.58}$$

where $PV = RT$, and the universal gas constant $R = 8.3143$ J mole^{-1} K^{-1}. $R = c_p - c_v$, therefore

$$c_v = c_p - R = R\left(\frac{c_p}{R} - 1\right) = R\left(\frac{c_p}{c_p - c_v} - 1\right) = R\left(\frac{\gamma}{\gamma - 1} - 1\right) = R\left(\frac{1}{\gamma - 1}\right). \tag{2.59}$$

(2.58) and (2.59) give

$$\begin{aligned}\frac{\Delta S}{R} &= \frac{1}{\gamma-1}\int_{T_0}^{T_1} \frac{\mathrm{d}T}{T} + \int_{V_0}^{V_1} \frac{\mathrm{d}V}{V} = \frac{1}{\gamma-1}\ln\left(\frac{T_S}{T_0}\right) + \ln\left(\frac{V_S}{V_0}\right) \\ &= \ln\left[\left(\frac{T_S}{T_0}\right)^{\frac{1}{\gamma-1}}\left(\frac{V_S}{V_0}\right)\right] = \ln\left[\left(\frac{T_S}{T_0}\right)^{\frac{1}{\gamma-1}}\left(\frac{\rho_0}{\rho_S}\right)\right].\end{aligned} \tag{2.60}$$

(2.60), (2.25), and (2.18) give

$$\frac{\Delta S}{R} = \ln\left\{\left[\frac{(2\gamma M_S^2 - \gamma + 1)\left((\gamma - 1)M_S^2 + 2\right)}{(\gamma + 1)^2 M_S^2}\right]^{\frac{1}{\gamma-1}}\left[\frac{(\gamma - 1)M_S^2 + 2}{(\gamma + 1)M_S^2}\right]\right\}$$

$$= \ln\left\{\left[\frac{2\gamma M_S^2 - \gamma + 1}{\gamma + 1}\right]^{\frac{1}{\gamma-1}}\left[\frac{(\gamma - 1)\,M_S^2 + 2}{(\gamma + 1)\,M_S^2}\right]^{\frac{\gamma}{\gamma-1}}\right\}. \tag{2.61}$$

For $\gamma = 1.4$ and $R = 8.3143$ J mole^{-1} K^{-1}, the change of entropy across a shock is

$$\Delta S = 8.3143 \ln\left\{\frac{\left(7M_S^2 - 1\right)^{2.5}}{46656}\left(\frac{M_S^2 + 5}{M_S^2}\right)^{3.5}\right\} \text{ J mole}^{-1}\text{ K}^{-1}. \tag{2.62}$$

2.4 Rankine–Hugoniot Spreadsheets

In order to make the above equations easy to use, they have been incorporated into a set of Excel© spreadsheets. The first spreadsheet requires the input of the shock Mach number, M_S, and calculates the changes of all the physical properties across the shock in terms of the ambient properties, and in SI and Imperial units. The conversions to SI and Imperial units are made assuming an ambient atmosphere at Normal Temperature and Pressure (NTP), viz. 15 °C (288.16 K) and 101.325 kPa (14.696 psi).

Illustrated below is the output for an input shock Mach number of 1.5.

	Hydrostatic Overpressure (OP)			Dynamic Pressure (DP)		
Ms	Atm	psi	kPa	Atm	psi	kPa
1.5	1.458333	21.43167	147.7656	0.628592	9.237787	63.69208

Total Overpressure (TOP)			Reflected Overpressure (ROP)		
Atm	psi	kPa	Atm	psi	kPa
2.146445	31.54415	217.4885	4.425287	65.03402	448.3922

Density (D)			Flow Speed (U)			
D/Do	lb/ft^3	kg/m^3	U/Ao	U/A	ft/s	m/s
1.862069	0.142448	2.281034	0.694444	0.604387	775.6097	236.4089

Sound Speed (A)			Temperature (T)		Inv. Press	Reflect Sound Speed	
A/Ao	ft/s	m/s	T/To	T (C)	Po/P	ft/s	m/s
1.149007	1283.3	391.1552	1.320216	107.2735	0.40678	1447.214	441.1164

The other 25 spreadsheets allow the input of the change of any of the physical properties across the shock, in any unit system. For example, the spreadsheet shown below is that for an input value for the dynamic pressure of 15 psi.

	Hydrostatic Overpressure (OP)			Dynamic Pressure (DP)		
Ms	Atm	psi	kPa	Atm	psi	kPa
1.623125	1.906956	28.02463	193.2223	1.020686	**15**	103.421

Total Overpressure (TOP)			Reflected Overpressure (ROP)		
Atm	psi	kPa	Atm	psi	kPa
3.062135	45.00114	310.2709	6.263559	92.04926	634.6551

Density (D)			Flow Speed (U)			
D/Do	lb/ft^3	kg/m^3	U/Ao	U/A	ft/s	m/s
2.070487	0.158392	2.536347	0.839191	0.708236	937.2739	285.6848

Sound Speed (A)			Temperature (T)		Inv. Press	Reflect Sound Speed	
A/Ao	ft/s	m/s	T/To	T (C)	Po/P	ft/s	m/s
1.184904	1323.393	403.3755	1.403996	131.4156	0.344002	1524.991	464.8233

The spreadsheets can be run on any PC, including laptops and palm computers. In calculating the total overpressure, allowance is made for both supersonic and subsonic flow behind the shock. Warnings are presented if the input values exceed the range of validity. The spreadsheets are demonstrated in www.blastanalysis.com, and information is provided on procedures for obtaining a license for their use.

2.5 Measurement Techniques

In order to use the Rankine–Hugoniot relationships to determine the values of the physical properties immediately behind the primary shock of a blast wave it is necessary to measure the Mach number, and therefore the velocity, of the shock. This entails measuring the time-of-arrival of the shock at a series of radial distances from the center of the explosion. Fortunately, the measurement of time of arrival of the primary shock is probably the most accurate measurement that can be made of any of the properties of a blast wave.

The time of arrival of the primary shock can be measured most simply by an array of detectors, sometimes known as ABTOADS (Air Blast Time-Of-Arrival Detectors). The detectors can be triggered by any of the physical properties that rise almost instantaneously at the shock front, particularly side-on and face-on pressure. Temperature sensors can be used, but must be protected from radiant heat from the luminous initial shock and the subsequent fireball. The detectors must have a rapid response, but do not need to be calibrated to relate the strength of the signal to the magnitude of the pressure or other change of physical property at the shock front. The signals from the detectors can be transmitted via a single cable to a recording device to produce a time-comb that can be related to the accurately surveyed positions of the detectors relative to the center of the explosion.

An example of the way in which this technique has been used to determine the physical properties immediately behind the primary shock of a centered explosion is described in Dewey (2005), and in Sect. 3.1 of Dewey (2015).

The method that has been used most extensively to measure the velocity of the primary shocks from centered explosions has been high-speed photography of the refractive image resulting from the distortion of the photographic background produced by the large gradient of the refractive index of the ambient atmosphere at the shock front as a result of the rapid change of the gas density. An example of the application of this method is described in Kleine et al. (2003). The technique is also described in Dewey (1997, 2015)

Acknowledgment The assistance of A. A. van Netten in making the numerical simulations for Figs. 2.3 and 2.4, using the AWAF code, is gratefully acknowledged.

References

Dewey, J. M. (1997). Shock waves from explosions, Chap. 16. In S. F. Ray (Ed.), *High speed photography and photonics*. Oxford: Focal Press.

Dewey, J. M. (2005). The TNT equivalence of an optimum propane–oxygen mixture. *Journal of Physics D: Applied Physics, 38*, 4245–4251.

Dewey, J. M. (2015). Measurement of the physical properties of blast waves. In O. Igra & F. Seiler (Eds.), *Experimental methods of shock wave research* (pp. 53–86). Cham: Springer.

Hugoniot, P. H. (1887). Mémoire sur la propagation du mouvement dans les corps et ples spécialement dans les gaz parfaits, 1e Partie. *Journal de l'École Polytechnique (Paris), 57*, 3–97.

Hugoniot, P. H. (1889). Mémoire sur la propagation du mouvement dans les corps et plus spécialement dans les gaz parfaits, 2e Partie. *Journal de l'École Polytechnique (Paris), 58*, 1–125.

Kleine, H., Dewey, J. M., Ohashi, K., Mizukaki, T., & Takayama, K. (2003). Studies of the TNT equivalence of silver azide charges. *Shock Waves, 13*, 123–138.

Liepmann, H. W., & Roshko, A. (1957). *Elements of gas dynamics*. Galcit aeronautical series. New York: John Wiley & Sons.

Prandtl, L., & Tietjens, O. G.. (1934). Fundamentals of hydro- and aerodynamics, Eng. Soc. Monographs, reprinted by Dover, New York, NY, 1957.

Rankine, W. J. M. (1870a). On the thermodynamic theory of waves of finite longitudinal disturbance, (read 16 Dec., 1869). *Philosophical Transactions of the Royal Society of London, 160*, 277–286.

Rankine, W. J. M. (1870b). Supplement to "On the thermodynamic theory of waves of finite longitudinal disturbance". *Philosophical Transactions of the Royal Society of London, 160*, 287–288.

Chapter 3
The Friedlander Equations

John M. Dewey

3.1 Introduction

The Friedlander equation, used to describe the pressure-time history of a blast wave, was first introduced in a paper by Friedlander (1946) that describes the analytical solutions of sound pulses diffracted by a semi-infinite plate. Friedlander was renowned for his erudicity and conciseness of presentation, and offers no explanation about the origin of the equation, or its possible association with blast waves. The equation was, in fact, first presented in a paper for the Civil Defence Research Committee of the British Ministry of Home Security by Sir Geoffrey (G. I.) Taylor (1941), albeit with distance rather than time as the independent variable. In the paper Taylor twice refers to analyses carried out by his research student, Mr Friedlander, but does not directly associate him with the equation that now bears his name.

It seems highly probable, therefore, that the equation presented by Friedlander was suggested to him by Taylor, and should probably be described as the Taylor equation. However, Sir Geoffrey Taylor had many physical phenomena named after him in his long and illustrious career, and it seems unlikely and he would begrudge the honour accorded to one of his students.

J.M. Dewey (✉)
Department of Physics and Astronomy, University of Victoria, Victoria, BC, Canada
e-mail: jdewey@uvic.ca

I. Sochet (ed.), *Blast Effects*, Shock Wave and High Pressure Phenomena,
https://doi.org/10.1007/978-3-319-70831-7_3

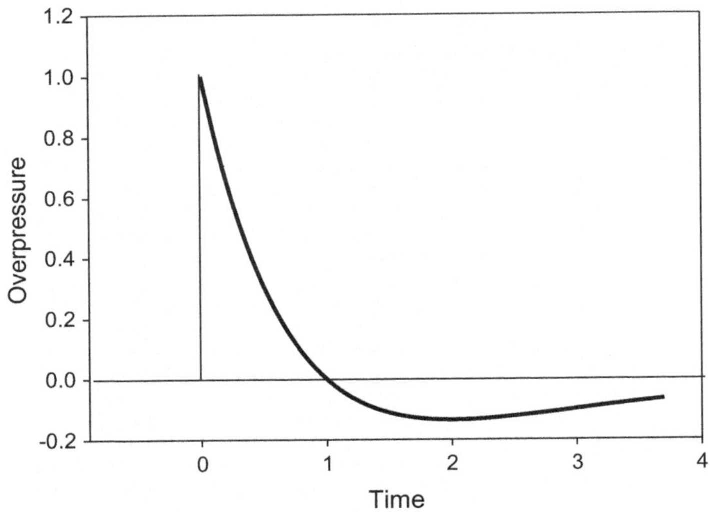

Fig. 3.1 The shape of the pressure pulse described by the Friedlander equation (3.1), with a peak overpressure of one unit, and a positive duration of one unit

3.2 The Friedlander Equation

The pressure-time history presented by Friedlander (1946) has the following form:

$$\mathrm{OP} = \mathrm{OP_s} e^{-t/t^+} \left(1 - t/t^+\right), \tag{3.1}$$

where OP is the hydrostatic overpressure, i.e. the pressure above the ambient value, $\mathrm{OP_s}$ is the overpressure immediately behind the shock front and t^+ is the duration of the positive phase, the period when the pressure is above the ambient value. The shape of the pressure pulse described by (3.1) is illustrated in Fig. 3.1. The similarity to the pressure-time history of a blast wave is immediately obvious, although for blast waves from explosive sources with a finite initial volume a second shock is produced that typically arrives in the region of the minimum pressure close to the time of 2 in Fig. 3.1.

In the 1940s the availability of piezo-electric transducers and associated amplifiers made it possible for the first to measure the hydrostatic pressure-time history in a blast wave. It soon became apparent that many of these time histories were excellently described by the Friedlander equation (3.1). The only requirement was to measure the peak overpressure, $\mathrm{OP_s}$, and the positive duration, t^+, from the recorded signal and insert them in (3.1) to obtain a complete time-history that could be compared with the measured result.

The excellent way in which the Friedlander equation is able to describe the time-histories of many blast wave properties is illustrated in Figs. 3.2, 3.3 and 3.4. These particular records are chosen for this illustration because they constitute a unique

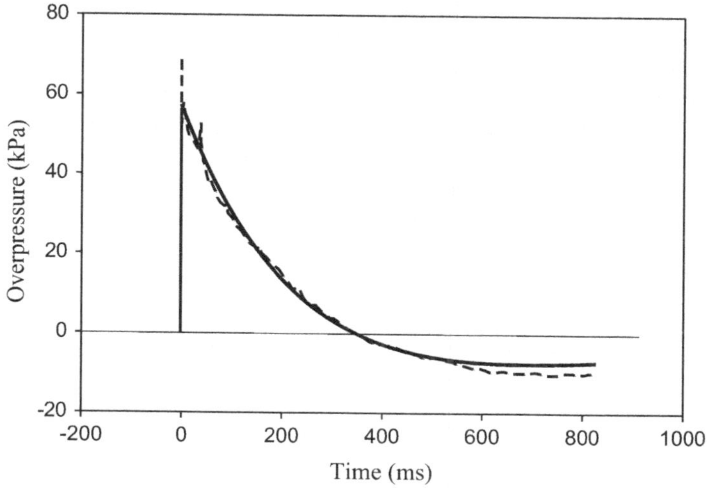

Fig. 3.2 Measured pressure time history (dashed curve) at a distance of 523.5 m from the centre of a surface-burst, hemispherical, 2.205 ktonne ammonium nitrate–fuel oil explosion (Slater et al. 1995). The black curve is the least squares fit of those data to the Friedlander equation (3.1)

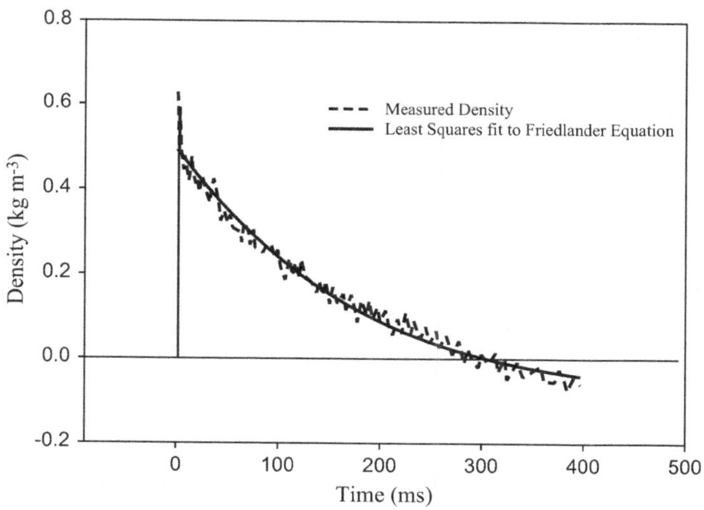

Fig. 3.3 Density time-history measured by Slater et al. (1995) using a β absorption gauge. The black curve is the least squares fit of these data to the Friedlander equation (3.1)

set of data, in that the time-histories of hydrostatic pressure, density and total (pitot) pressure were simultaneously measured at the same location. The gauge signals were digitised and least-squares fitted to the Friedlander equation (3.1).

The expanding and decaying spherical shock of the blast wave from a centred explosion leaves the air in a state of radially decreasing entropy. As a result, the

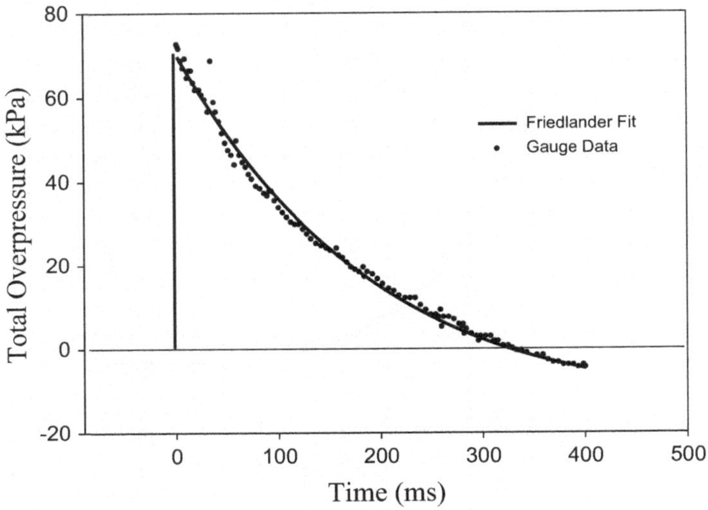

Fig. 3.4 Total (Pitot) pressure (dots) measured by Slater et al. (1995). The black curve is the least squares fit of these data to the Friedlander equation (3.1)

blast wave flow passing a fixed location is non-isentropic and the simple isentropic thermodynamic relationships cannot be used to relate physical properties measured at that location. This means, for example, that the hydrostatic pressure time-history measured at a fixed point in a blast wave cannot be used to determine the time histories of other physical properties such as density, temperature and particle velocity. To fully describe all the physical properties at a fixed point in a blast wave it is necessary to independently measure at least three of those properties. In order to achieve this, the Defence Research Establishment Suffield (DRES), Alberta, Canada, developed a blast measurement station that independently measured the time-histories of hydrostatic and total (pitot) pressures, and the density, at the same location. The hydrostatic and total pressures were measured by piezo-electric transducers mounted flush and face on to the blast flow, and the density was measured using β-ray absorption (Dewey and Anson 1963). DRES blast stations, described in more detail by Slater et al. (1995) and Dewey (2016), were used at several locations to monitor the blast wave produced by a surface-burst, hemispherical 2.205 kt ammonium nitrate-fuel oil explosion (MINOR UNCLE). The station that recorded the data shown in the figures was at a distance where the nominal peak hydrostatic overpressure was about 0.5 atm.

The measurements of hydrostatic and total pressures were used by Slater et al. to determine the dynamic pressure time-history, as described in Sect. 3.8 of Dewey (2016). For blast wave applications the dynamic pressure is defined as $P_{\mathrm{D}} = 1\backslash 2\ \rho\ u\ |u|$, where ρ is the gas density and u the particle velocity, rather than the more common definition, $1\backslash 2\ \rho\ u^2$, so that it is a vector rather than a scalar property of the flow. The dynamic pressure results were least-squares fitted to the Friedlander equation with the result shown in Fig. 3.5. Again, the fit was excellent,

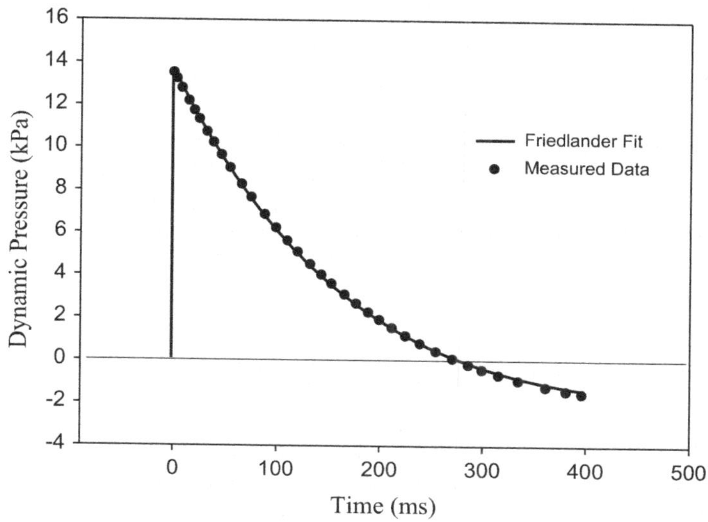

Fig. 3.5 Dynamic pressure (dots) derived from the results of Slater et al. (1995). The black curve is the least squares fit of these data to the Friedlander equation (3.1)

although this was not necessarily expected because the dynamic pressure expressed as 1\2 $(F)_{D}$ $(F)_{U}$^2, where $(F)_{D}$ and $(F)_{U}$ are the Friedlander time-histories of density and particle velocity, respectively, does not have the same algebraic form as the Friedlander equation.

The excellent way in which the Friedlander equation is able to describe the time histories of the hydrostatic, total and dynamic pressures, and the density as illustrated in Figs. 3.2, 3.3, 3.4 and 3.5 is typical for blast waves from most solid explosives, such as TNT, pentolite, C4 and ANFO, with peak hydrostatic overpressures less than about 1 atm. This is not the case for all the physical properties of a blast wave. One exception is the time history of particle velocity from TNT explosions. TNT is an oxygen deficient explosive, and as a result the detonation products continue to burn as they turbulently mix with the ambient air. This phenomenon has little effect on the hydrostatic pressure and density of the blast wave, but it does cause a continuing outward flow such that the particle velocity does not reverse at the end of the positive phase of pressure. This phenomenon is described and explained by Dewey (1964), and an alternate to the Friedlander equation is presented to describe the particle velocity time-histories in blast waves from TNT explosions.

Another exception is temperature. The expanding and decaying shock from a centred explosion, in addition to its effect on entropy, as described above, also leaves the air in a state of radially decreasing temperature and radially increasing density. After the passage of a blast wave the pressure rapidly returns to the ambient value because any gradient in pressure is relieved at the speed of sound. The gradients in temperature and density can only be relieved by particle flow, and so after the passage of the blast wave the air temperature is higher than the ambient value, and the density is less than its ambient value. This means that the temperature measured at a fixed location does not decay below the ambient value, and after the initial decay will show a significant increase. The resulting time-history of temperature is not described by the Friedlander equation.

The corresponding gradient in density has little effect on the shape of its time-history during the passage of the blast wave. Subsequently, however, the buoyancy effect of the density gradient causes an extended upward and inward flow that is responsible for the long cylindrical stem of the mushroom cloud produced by the explosion.

3.3 Properties of the Friedlander Equation

The excellent way in which the Friedlander equation describes the time histories of the physical properties of blast waves over a range of distances from the centre of the explosion means that a number of its unique properties can be used to assist the analysis of these blast waves.

The most notable feature of the Friedlander equation is that it includes only two independent parameters, viz. the peak value of the physical property, e.g. $\mathrm{OP_S}$ in the case of pressure, and t^+, the positive duration. Both of these properties can be measured with relative ease from a pressure-time record, and used in the Friedlander equation to describe the shape of the wave and the positive phase impulse, I_+, as will be described below. Many of the properties of the Friedlander equation were first developed and described by Thornhill (1959).

In many cases it may not be possible to accurately identify the peak value or the positive duration from a gauge record due to noise in the signal. This noise may arise from external sources, but is most commonly caused by the mechanical inertia of the pressure transducer and the electrical inertia of the associated electronic amplifiers and recorder. These inertias mean that the transducer is unable to accurately follow the very rapid change of pressure and other physical properties at a shock front, but then overshoots causing ringing in the early part of the pressure signal.

To overcome this problem the pressure values in the first one-third of the signal can be plotted on semi-log paper. A straight line drawn through these points can be extrapolated to the time of arrival of the shock to give the best estimate of the peak pressure. Alternatively, the log of the pressures can be least-squared fitted to a straight line, viz.

$$\ln(\mathrm{OP}) = A + Bt, \tag{3.2}$$

where OP is the recorded overpressure, t the time measured from the time of arrival of the shock, and A and B are the fitted coefficients. At $t = 0$ the exponential of A is the best estimate of the peak overpressure, $\mathrm{OP_S}$. Also it may be noted that the second fitted coefficient $B = \mathrm{d}(\ln\mathrm{OP})/\mathrm{d}t$, and this may be used as follows.

Close to the shock front $t \rightarrow 0$, and the Friedlander equation may be approximated as

$$\mathrm{OP} = \mathrm{OP_S} e^{-\frac{t}{t^+}}. \tag{3.3}$$

Therefore,

$$\ln \mathrm{OP} = \ln \mathrm{OP_S} - \frac{t}{t^+}. \tag{3.4}$$

Differentiating (3.4) gives

$$\frac{\mathrm{d}\,(\ln \mathrm{OP})}{\mathrm{d}t} = -\frac{1}{t^+}. \tag{3.5}$$

This means that $t^+ \approx -1/\frac{\mathrm{d}(\ln \mathrm{OP})}{\mathrm{d}t} = -1/B$, where B is the second fitted coefficient in (3.2). In other words, for a region in which the Friedlander equation is known to be valid, using only a limited number of the pressure measurements immediately behind the shock is sufficient to determine the peak overpressure, $\mathrm{OP_S}$, and to obtain a good estimate of the positive duration. This can be useful if the pressure-time record is truncated by a gauge failure, for example, or the signal is disturbed by the arrival of the shock reflected from a nearby structure.

The positive phase duration, t^+, also is not always easy to read accurately from a pressure-time record. The signal may be noisy in this region, and the second shock may arrive in the positive phase, something that can occur with TNT and propane/oxygen explosions, and there is no clear intersection of the record with the $P = 0$ axis. Also, in the early days of blast wave measurements, many of the piezoelectric transducers, particularly quartz transducers, were temperature-sensitive so that as the gauge was heated by the temperature change in the blast wave the calibration gradually changed and caused something called "base line shift", i.e. the exact position of the $P = 0$ axis was not known. It was found that if the last one-third of the points in the positive phase of the pressure-time record were plotted on semi-log paper, the result was almost linear and a straight line could be drawn through the points to determine the intersection with the $P = 0$ axis and provide a value for t^+.

Thornhill (1959) suggested another solution for this problem. The relaxation time, t^*, of an exponential decay is defined as the time for the property to decay to $1/e$ of its peak value. Appling this to the Friedlander equation (3.1) gives

$$1 = e^{\left(1-\frac{t^*}{t^+}\right)}\left(1 - \frac{t^*}{t^+}\right), \tag{3.6}$$

and the solution of (3.6) is

$$t^+ = 2.31t^*. \tag{3.7}$$

The relaxation time, t^*, may be more easily read from a pressure-time record and can be used in (3.7) to estimate t^+ when there is no clear or accurate intersection of the signal with the $\mathrm{OP} = 0$ axis.

One of the most important pieces of information that is required from a pressure-time record is the positive phase impulse, I_+, the pressure integral over that period. For a record that is well described by the Friedlander equation

$$I_+ = \int_0^{t^+} \mathrm{OP_S} e^{-\frac{t}{t^+}} \left(1 - \frac{t}{t^+}\right) \mathrm{d}t = 0.368\mathrm{OP_S} t^+, \tag{3.8}$$

so that the impulse can be determined from the peak overpressure and the positive phase duration only.

Alternatively, using (3.7) and (3.8)

$$I_+ = 0.85\mathrm{OP_S} t^*, \tag{3.9}$$

and the impulse is obtained from the peak overpressure and the relaxation time.

There are some other interesting features of the Friedlander equation. Differentiating (3.1) and equating it to zero give

$$t_{\min} = 2t^+ = 4.6t^*, \tag{3.10}$$

where $t_{\min}$ is the time at which the pressure reaches its minimum value in the negative phase. Inserting this value in (3.1) gives the minimum value of the pressure in the negative phase

$$\mathrm{OP_{min}} = -\mathrm{OP_S} e^{-2} = -0.135\mathrm{OP_S}. \tag{3.11}$$

All of the above relationships have been validated by comparing them with many pressure-time records of blast waves in the region for which the Friedlander equation is valid.

Most of the above relationships were developed and used for the analysis of blast wave pressure-time records in the days of analogue recording and before the availability of digital computers. When digital computers became available it was possible to least-squares fit the pressure-time data to the Friedlander equation using non-linear regression programs. Such programs are available with most analytical mathematics applications, including, for example, Excel©. Least-squares fitting blast wave properties to the Friedlander equation immediately provides the peak value of the physical property and the positive duration, as the fitted coefficients. This is now the preferred procedure for deriving those values from a recorded time history.

Before making the least-squares fit, the data to be fitted must be selected from the total data recorded. It is recommended that the data to be fitted be initiated from slightly behind the arrival of the shock to eliminate the data during the inertially slowed rise time and subsequent overshoot and ringing when the gauge is clearly not accurately recording the pressure in the blast wave. The data is further truncated at the arrival of the second shock or close to the minimum of the pressure in the negative phase. This technique was used to make the least-squares fits shown in Fig. 3.2. For blast waves from solid explosives such as TNT, pentolite and ammonium-nitrate fuel-oil (ANFO) it was observed that the time histories of the physical properties were well described by the Friedlander equation only at relatively large distances from the centre of the explosion, typically when the peak hydrostatic overpressure was less than about 1 atm. As a result, the Friedlander equation was modified to a form that would accurately describe the shape of the blast waves at much shorter radial distances, up to the boundary of the detonation products, which corresponds to a peak hydrostatic overpressure of about 7 atm.

3.4 The Modified Friedlander Equation

The Friedlander equation was initially modified by adding an additional factor, α, to the exponential coefficient in (3.1), which then becomes

$$\mathrm{OP} = \mathrm{OP_s} e^{-\propto t/t^+} \left(1 - {t}/{t^+}\right). \tag{3.12}$$

Since α and t^+ are constants, (3.12) may also be written as

$$\mathrm{OP} = \mathrm{OP_S} e^{-\beta t} \left(1 - {t}/{t^+}\right). \tag{3.13}$$

If the same set of pressure-time data is fitted to both (3.12) and (3.13) it will be found that the fitted coefficients $\mathrm{OP_S}$ and t^+ will be similar but not identical for the two equations. Also α/t^+ from Eq. (3.12) will not exactly equal β from Eq. (3.13). Comparing the goodness-of-fit measures for the two least-squares fits shows that the fit to (3.13) is slightly better than that to (3.12), and it is therefore recommended that (3.13) be used as the so-called modified Friedlander equation.

Figure 3.6 shows an attempt to fit the Friedlander equation to the pressure time-history at a distance of 1.3 m from the free-field explosion of 1 kg TNT where the peak hydrostatic overpressure was approximately 3.7 atm. The data were derived from the *AirBlast*[1] database. It can be seen that the Friedlander equation is unable to describe the pressure-time history, and fails to identify the peak overpressure and positive duration. In contrast, the modified Friedlander equation (3.13) provides an excellent description of the time history, and correctly identifies the peak overpressure and the positive duration.

The modified Friedlander equation (3.13) has been shown to provide a good description of overpressure time-histories up to a peak hydrostatic overpressure level of about 7 atm, for solid explosives such as TNT, ANFO and pentolite, and a stoichiometric mixture of propane and oxygen. It also provides a good description at the low overpressures where the unmodified Friedlander equation (3.1) is also

[1]*AirBlast* is an interactive database of blast wave properties provided by Dewey McMillin & Associates www.blastanalysis.com.

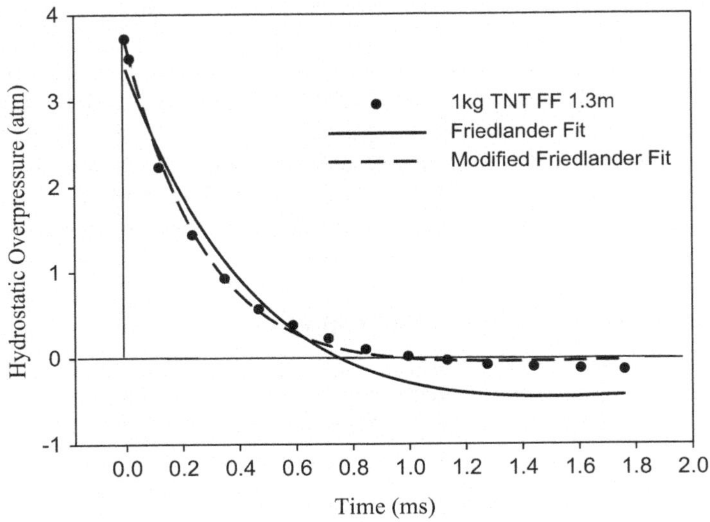

Fig. 3.6 The data points show the hydrostatic overpressure time history derived from *AirBlast* at 1.3 m from a free-field 1 kg TNT explosion. The black curve is the Friedlander equation (3.1) fit to those data. The dashed curve is the modified Friedlander equation (3.13) fit to the same data

valid. In that same wide range of distances from the centre of an explosion, from close to the contact surface with the detonation products to the point where the shock wave is beginning to degenerate into a sound wave, (3.13) also describes the time histories of density, total and dynamic pressures, and particle velocity, with the exception described in Sect. 3.3 for TNT.

3.5 Properties of the Modified Friedlander Equation

The modified Friedlander equation is a useful tool in the analysis of most blast waves. (3.13) can be least-squares fitted, using non-linear regression, to the measured time-histories of hydrostatic overpressure, overdensity and particle velocity. The resulting fitted coefficients, OP_S and t^+, are usually the best estimates of the peak value of the property immediately behind the shock, and the positive duration. This is particularly useful when analysing a noisy signal.

The nonlinear regression analysis requires the initial input of estimates for the fitted coefficients OP_S, t^+ and β. These estimates can be obtained using the same techniques as those described in Sect. 3.3 and (3.2)–(3.5).

The positive phase impulse, I_+, is given by

$$I_+ = \int_0^{t^+} \mathrm{OP_S} e^{-\beta t} \left(1 - {}^{t}\!/_{t^+}\right) \mathrm{d}t = \frac{\mathrm{OP_S}}{\beta^2 t^+} \left(e^{-\beta t^+} + \beta t^+ - 1\right). \tag{3.14}$$

The impulse is therefore easily calculated from the three fitted coefficients, $\mathrm{OP_S}$, β and t^+.

Integrating the modified Friedlander equation from $t = 0$ to ∞ gives

$$I_{\mathrm{Tot}} = \int_0^{\infty} \mathrm{OP_S} e^{-\beta t} \left(1 - {}^{t}\!/_{t^+}\right) \mathrm{d}t = \mathrm{OP_S} \left(\frac{t^+ - 1}{\beta t^+}\right). \tag{3.15}$$

Unlike the total integral of the unmodified Friedlander equation, the total integral of the modified equation has a finite value. This is not unexpected because the total integral of the energy in a blast wave would be expected to have a value equal to the energy released from the explosion.

The minimum overpressure of the modified Friedlander equation occurs at a time of

$$t_{\mathrm{min}} = t^+ \left(1 + {}^{1}\!/_{\beta}\right). \tag{3.16}$$

Inserting this value in (3.13) gives the minimum overpressure, although for most explosives that have been studied the second shock arrives before this time.

The modified Friedlander equation can be used to interpolate time histories at distances between those at which measurements were made. The equation is least-squares fitted to a series of time-histories at different distances from an explosion. The fitted coefficients, $\mathrm{OP_S}$, β and t^+, can then be plotted against radius, or fitted to functions, $F_{\mathrm{OP}}(R)$, $F_{\beta}(R)$ and $F_t{}^+(R)$. Inserting intermediate radii into these functions provides values of $\mathrm{OP_S}$, β and t^+ that can be used in the modified Friedlander equation to determine the time histories at the intermediate radii.

3.6 Wave Profiles

The physical properties of blast waves are usually viewed as time histories. This is because the properties are most easily measured at fixed locations as the blast wave traverses those locations. However, if the blast wave properties are derived from a numerical simulation, or a combination of the numerical simulation and measurement, such as the piston path method, the properties can also be viewed as functions of distance at fixed times, viz. wave profiles. It has been found that for most centred explosions the wave profiles of the physical properties are also well described by the modified or the unmodified Friedlander equation where radius is now the independent variable.

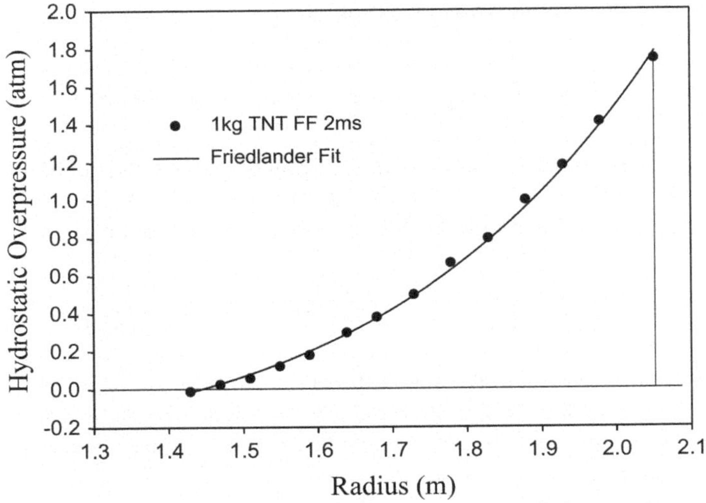

Fig. 3.7 Hydrostatic overpressure wave profile at 2 ms after a free-field detonation of 1 kg TNT. The points are derived from *AirBlast*, and the black curve is the least squares fit to the Friedlander equation (3.17)

The Friedlander equation now has the form

$$\mathrm{OP} = \mathrm{OP_S} e^{-\frac{r}{r^+}} \left(1 - {}^{1}/_{r^+}\right), \tag{3.17}$$

where $r = R_S - R$, R_S is the radius of the primary shock, R the radius of any other point in the wave profile and r^+ is the distance behind the primary shock where the overpressure first returns to its ambient value. The modified equation has a similar form but with an exponential coefficient of $-\beta r$. An example of a least-squares fit of the Friedlander equation to the profile of a blast wave from a free-field TNT explosion is shown in Fig. 3.7.

It will be noted that the peak overpressure of the wave profile illustrated in Fig. 3.7 is above 1 atm, the upper limit of the Friedlander equation for time histories. In the case of wave profiles, the Friedlander equation is valid for peak overpressures above 1 atm, but not below. This is illustrated in Fig. 3.8, in which the modified and unmodified Friedlander equations have been least-squares fitted to the wave profile of a TNT explosion for which the peak hydrostatic overpressure was approximately 0.3 atm.

3.7 Valid Range of the Friedlander Equations

For most solid explosives, and for stoichiometric propane-oxygen explosions, the time histories and wave profiles of most of the physical properties of the blast waves

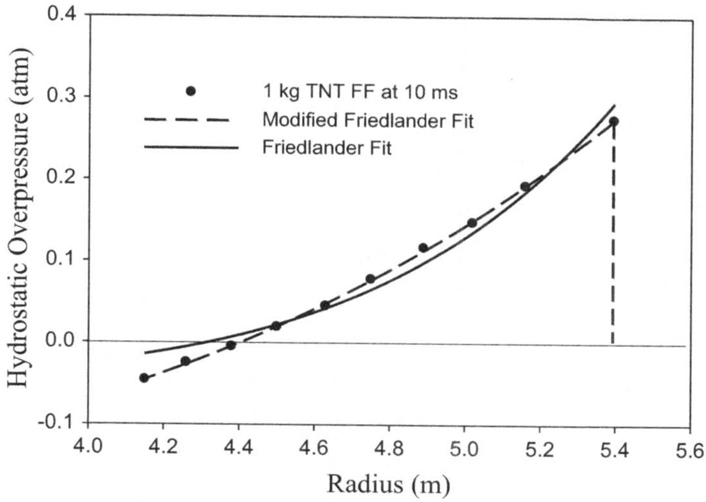

Fig. 3.8 Hydrostatic overpressure wave profile at 10 ms after a free-field detonation of 1 kg TNT. The points are derived from *AirBlast*. The solid and dashed curves are, respectively, the least squares fits to the Friedlander and modified Friedlander equations

are well described by the modified Friedlander equation over a range of distances from close to the contact zone of the detonation products to the point where the blast wave is degenerating into a sound wave. There is a much smaller range of distances in which the Friedlander equation is a valid descriptor.

To determine the region in which the Friedlander equation is a valid descriptor of the blast wave properties, the following procedure can be used. In the region in which the Friedlander equation is valid, the exponential coefficient in the modified Friedlander equation, β, is approximately equal to the exponential coefficient in the Friedlander equation, $1/t^+$.

This is illustrated in Fig. 3.9. The hydrostatic pressure-time histories of the blast wave produced by a 1 kg TNT free-field explosion were least-squares fitted to the modified Friedlander equation and the fitted values of t^+ and $1/\beta$ are plotted versus the peak hydrostatic overpressure. The peak overpressure has used as the independent variable, rather than radial distance, so that the results are independent of charge mass. It can be seen that the region in which t^+ is approximately equal to $1/\beta$ extends from a peak hydrostatic overpressure of approximately 1 atm to approximately 0.5 atm. It appears that below a peak hydrostatic overpressure of 0.5 atm the two curves may begin to separate, and so it may be necessary to use the modified Friedlander equation at very low pressures. This is a subject for further evaluation.

Applying the same analysis to the wave profiles gives the result shown in Fig. 3.10.

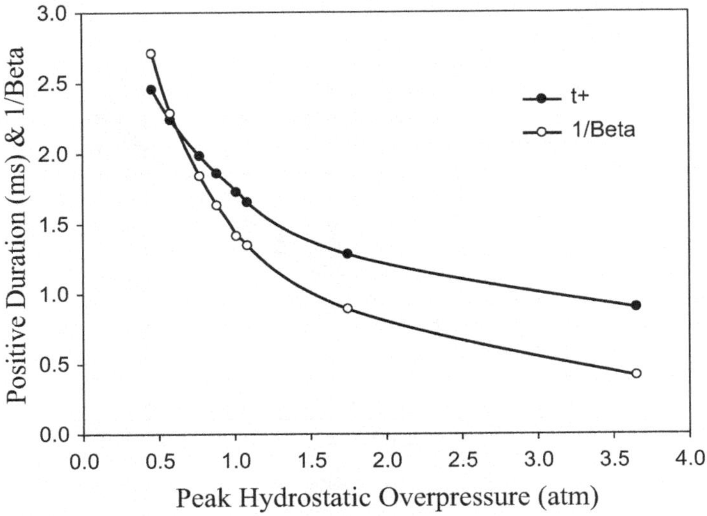

Fig. 3.9 The values of t^+ and $1/\beta$ obtained from the least-squares fits of the modified Friedlander equation to a series of hydrostatic overpressure time-histories for the free-field explosion of 1 kg TNT, plotted versus peak hydrostatic overpressure

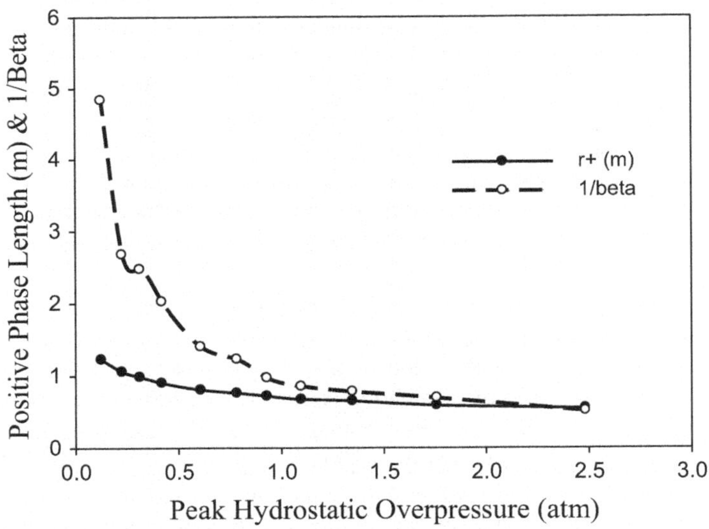

Fig. 3.10 A comparison of the positive phase length (solid points) with the inverse of the exponential coefficient (open points), of least-squares fits to the modified Friedlander equation of the wave profiles of the blast wave from a 1 kg free-field TNT explosion, plotted versus the peak hydrostatic pressures

It can be seen that in contrast to the similar comparison for the time histories, the positive phase lengths and the inverses of the exponential coefficient are very similar over a range of the peak hydrostatic overpressures from 2.5 atm to 1 atm. This indicates that the unmodified Friedlander equation could be used to describe the wave profiles in that region. For peak overpressures less than 1 atm the curves rapidly diverged, indicating the need to use the modified Friedlander equation to describe wave profiles in that region.

The lower curve in Fig. 3.10 also demonstrates that there is very little change in the length of the blast wave from the time at which it emerges from the detonation products, when the peak hydrostatic overpressure is approximately 2.5 atm, to the point at which the peak overpressure has decayed to about 0.1 atm.

The same analysis has been applied to the blast wave generated by a stoichiometric propane/oxygen explosion. The physical properties of the explosion were obtained from the analysis of measurements made of the blast wave generated by a 20 ton stoichiometric propane/oxygen explosion described by Dewey (2005) and Dewey and Dewey (2014). The time histories and wave profiles were derived from an interactive database obtained from the above analyses (Dewey 2016).

Using the interactive database, a series of hydrostatic overpressure time-histories were obtained for a range of distances from a 1 kg stoichiometric propane explosion, and these were least-squares fitted to the modified Friedlander equation. The resulting values of the positive duration and the inverse of the exponential coefficient (β) are plotted versus peak hydrostatic overpressure in Fig. 3.11.

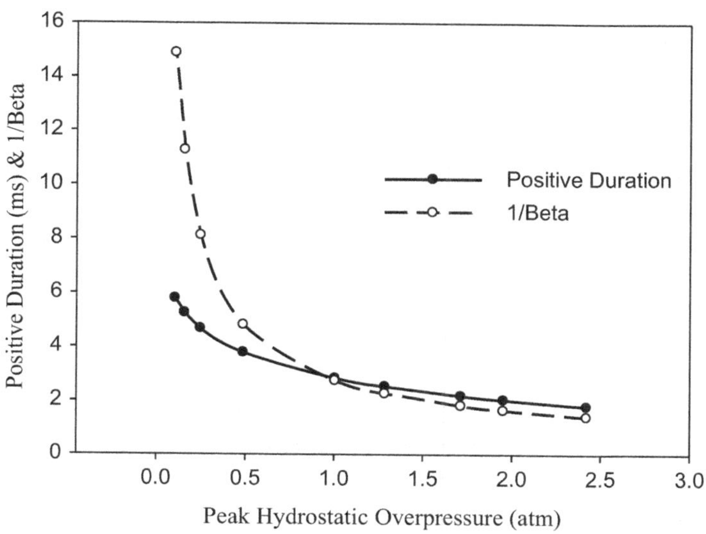

Fig. 3.11 Positive duration (solid points) and the inverse of the exponential coefficient (open points) plotted versus peak hydrostatic overpressure for a series of hydrostatic pressure time-histories of the blast wave generated by the explosion of 1 kg propane in a stoichiometric mixture with oxygen, derived from a propane/oxygen interactive database

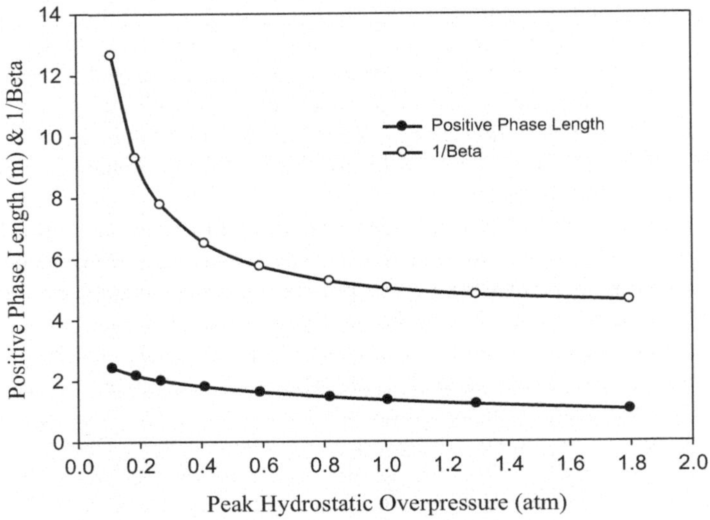

Fig. 3.12 Positive phase length (solid points) and the inverse of the exponential coefficient (open points) plotted versus peak hydrostatic overpressure for a series of hydrostatic pressure wave profiles for the blast wave generated by the explosion of 1 kg propane in a stoichiometric mixture with oxygen, derived from a propane/oxygen interactive database

It can be seen that, in contrast to the results for solid explosives, the unmodified Friedlander equation can be used to describe the pressure time-histories from the high overpressure region close to where the blast wave is emerging from the detonation products, to the point where the peak hydrostatic overpressure has fallen to about 1 atm. At greater distances it is necessary to use the modified Friedlander equation to describe the time histories.

A similar analysis of the wave profiles produced by a propane-oxygen explosion gives the results shown in Fig. 3.12.

From this analysis, it appears that for a propane explosion there is no region in which the hydrostatic overpressure wave profiles can be described by the unmodified Friedlander equation. For the range of overpressures shown in Fig. 3.12 the wave profiles are perfectly described by the modified Friedlander equation.

3.8 Origin of the Friedlander Equation

Several authors, including Thornhill (1959), noting that the simple Friedlander equation, with only two fitted coefficients, is such an excellent descriptor of the blast waves produced by a variety of free-field and surface-burst explosives, have asked if there might not be a more fundamental relationship between the physical processes of an explosion and this equation, i.e. that it is not merely an empirical relationship.

Taylor (1946) suggested that any centred blast wave can be produced by a spherical piston with an appropriate radius-time trajectory. That trajectory can be determined experimentally by high speed photogrammetry of a flow tracer, i.e. a smoke puff or trail established close to an explosive source shortly before initiation. This technique has been used extensively during the past 50 years to reconstruct the blast wave properties produced by high explosives ranging from a few kilograms to several kilotons. The physical properties obtained from such reconstructions are in excellent agreement with measurements using electronic transducers to record hydrostatic and total pressures, and density (Dewey 1997a, b, 2000, 2016).

Figure 3.13 shows the particle trajectory used to reconstruct the blast wave from a hemispherical surface-burst explosion of 2445 tons of ANFO (Operation MISERS GOLD).

The piston path points in Fig. 3.13 have been rotated and fitted by least squares to a form of the Friedlander equation, viz.

$$R = R_0 + R_{\max} e^{-\frac{t}{t^+}} \left(1 - \frac{t}{t^+}\right), \tag{3.18}$$

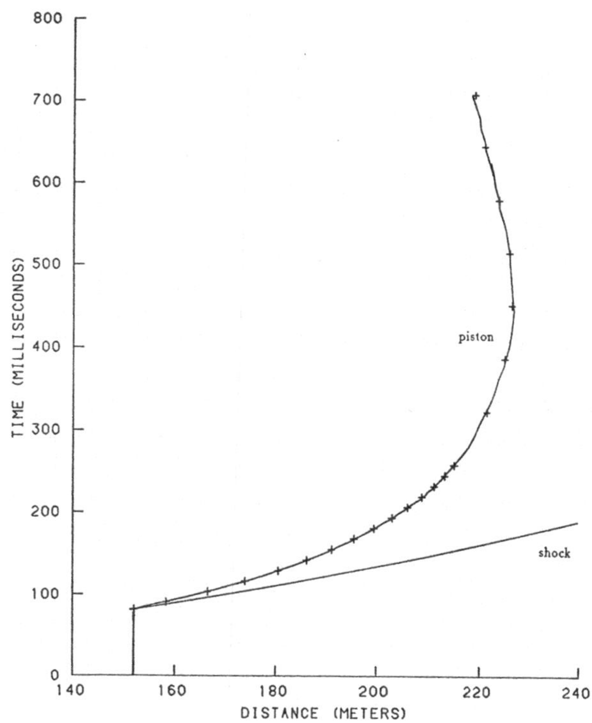

Fig. 3.13 The upper curve is the trajectory of a smoke tracer originally formed at about 150 m from the centre of a 2445 ton ANFO explosion. The points on the curve were measured and used as the spherical piston path to reconstruct the physical properties of the blast wave

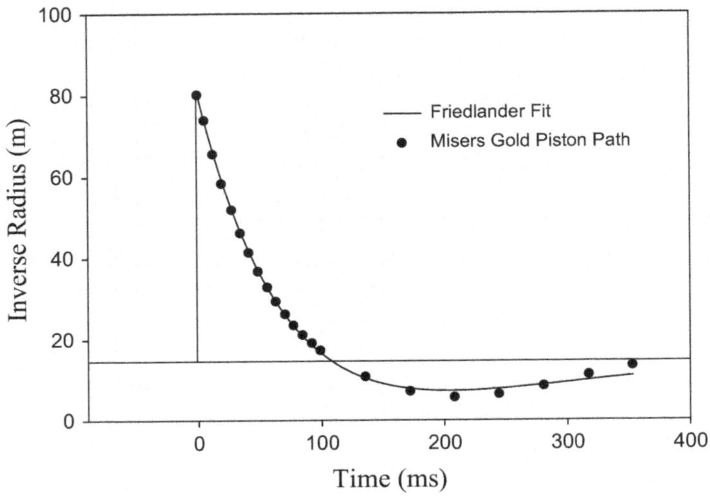

Fig. 3.14 Points from the spherical piston path of MISERS GOLD fitted by least squares to the Friedlander equation (3.18)

where $R_{\text{max}} = R - R_0$, R and t are the radius and time of the measured piston path, measured from the initial point of the piston path, and R_0 is an arbitrarily chosen radius. The resulting fit to this form of the Friedlander equation is shown in Fig. 3.14, and it is clear from this result that the spherical piston that drives a centred blast wave has the form of the Friedlander equation, and thus, it should not be unexpected that the resulting wave profiles and time histories also have that form. It has been shown that this form of the spherical piston path driving a blast wave is valid for solid explosives such as TNT, pentolite and ANFO, and for stoichiometric explosions of propane and oxygen.

In Sect. 3.1 it was suggested that the equation we now know as the Friedlander equation was possibly originated by Sir Geoffrey Taylor. With the evidence shown in Fig. 3.14, this seems to be even more probable because at the beginning of World War II, Prof G.I. Taylor was the principal scientific consultant to the British Ministry of Defence on the subject of explosions and blast waves in air (Batchelor 1996). One of the topics studied by a committee on which Taylor served was the trajectory of the contact surface between the detonation products and the air from TNT explosions.

Acknowledgement The author gratefully acknowledges Douglas McMillin, who first observed the Friedlander form of the spherical-piston path, and who developed the piston path method for the reconstruction of the physical properties of centred blast waves.

References

Batchelor, G. (1996). *The life and legacy of G.I. Taylor*. Cambridge: Cambridge University Press.

Dewey, J. M. (1964). The air velocity in blast waves from t.n.t. explosions. *Proceedings of the Royal Society A, 279*, 366–385.

Dewey, J. M. (2005). The TNT equivalence of an optimum propane-oxygen mixture. *Journal of Physics D: Applied Physics, 38*, 4245–4251.

Dewey, J. M. (1997a). Shock waves from explosions, Chap. 16. In S. F. Ray (Ed.), *High speed photography and photonics* (pp. 245–253). Oxford: Focal Press.

Dewey, J. M. (1997b). Explosive flows: Shock tubes and blast waves, Chap. 29. In W.-J. Yang (Ed.), *Handbook of flow visualization* (2nd ed.). New York: Hemisphere.

Dewey, J. M. (2000). Spherical expanding shocks (Blast waves). In *Handbook of shock waves* (Vol. 2, 13.1 ed., pp. 441–481). New York: Academic Press.

Dewey, J. M. (2016). Measurement of the physical properties of blast waves. In O. Igra & F. Seiler (Eds.), *Experimental methods of shock wave research*. Berlin: Springer.

Dewey, J. M. (2016). A user interface to provide the physical properties of blast waves from propane explosions. In *Proceedings of the 22nd international symposium on Military Aspects of Blast and Shock*, MABS22, Halifax, Canada.

Dewey, J. M., & Anson, W. A. (1963). A blast wave density gauge using beta radiation. *Journal of Scientific Instruments, 40*, 568–572.

Dewey, M. C., & Dewey, J. M. (2014). The physical properties of the blast wave produced by a stoichiometric propane/oxygen explosion. *Shock Waves, 24*, 593–601.

Friedlander, F. G. (1946). The diffraction of sound pulses. I. Diffraction by a semi-infinite plate. *Proceedings of the Royal Society of London A, 186*, 322–344.

Slater, J. E., Boechler, D. E., & Edgar, R. C. (1995). DRES measurement of free-field airblast. In *Minor Uncle Symposium Report*, Defense Nuclear Agency, POR 7453-4 (Vol. 4, 2, pp. 1–98).

Taylor, G. I. (1941). The propagation of blast waves over the ground. In G. K. Batchelor (Ed.), *G. I. Taylor scientific papers* (Vol. 3). Cambridge: Cambridge University Press.

Taylor, G. I. (1946). The air wave surrounding an expanding sphere. *Proceedings of the Royal Society of London A, 186*, 273–292.

Thornhill, C. K. (1959). *The shape of a spherical blast wave, Armament Research and Development Establishment (ARDE) Memo. (B) 41/59*. London: HMSO.

Chapter 4
Shock Wave Overpressure Measurement: Comparison Between Two Piezoelectric Sensor Materials

Pietro Tadini, Kevin Gault, and Isabelle Sochet

Extensive knowledge on shock waves has been developed and, nowadays, analytical, empirical, and numerical tools are available in the literature (Zel'dovich and Raizer 1966, 1967; Kinney and Graham 1985; Ben-Dor et al. 2001) to predict the shock behavior, as well as the effect of explosions on buildings (U.S. Department of the Army 1990; Remennikov 2003). Moreover, in recent years, a growing attention was given to the study of blast waves inside confined geometries, such as tunnels or industrial buildings (Schelinski-Glück 1993; Miura et al. 2004; Julien et al. 2016). Thus, different measurement techniques can be implemented for the experimental characterization of shock waves, such as photogrammetric, passive, and electronic methods (Settles 2012; Dewey 2001). In particular, the latter is one of the most simple and reliable approaches to measure the hydrostatic and total pressure of the blast. These transducers can be realized by strain, capacitance or piezoelectric sensors, finding a large use especially in lab-scale setups, due to their small size and reduced perturbation of the flow field. Several examples of piezoelectric transducers, described in detail with regard to their size and performance, are provided in Reisler et al. (1995). Such type of gauges need a pre-amplification device due to the small voltage change produced by the piezoelectric element when loaded by a blast wave. Then, the small current transmitted in the cable must be amplified by the acquisition system. In this kind of applications, the probe is subjected to both radiative and convective heating, thus requiring a piezoelectric material with very low sensitivity to temperature. In addition, the transducer must be properly isolated from its holding structure (e.g., table, rods, tripod) in order to limit spurious signals associated with the propagation of structural vibrations (Dewey 2016). Depending on the way of mounting, the hydrostatic or the stagnation pressure can be captured: the former by placing the gauge with its sensing surface parallel to the flow, whereas the

P. Tadini (✉) • K. Gault • I. Sochet
Laboratoire PRISME, INSA Centre Val de Loire, Bourges, France
e-mail: pietro.tadini@insa-cvl.fr

I. Sochet (ed.), *Blast Effects*, Shock Wave and High Pressure Phenomena,
https://doi.org/10.1007/978-3-319-70831-7_4

second perpendicular to the shock propagation direction, assembled in a cylindrical mount supported by a thin rod (e.g., Pitot tube) (Chue 1975; Josey et al. 2016). The dynamic pressure can then be estimated, by the Rayleigh supersonic Pitot formula (Dewey 2016), with these two independent measures achieved simultaneously in the same local area. In order to obtain a reliable pressure-time history, the high-frequency transducers must be accurately calibrated by a pressure-voltage relationship.

Alternatively, an even easier method, always by exploiting a piezoelectric high-frequency sensor, consists in the measuring of primary shock arrival time, hence allowing the estimation of wave velocity and Mach number, knowing the temperature of ambient air. Then, by means of Rankine-Hugoniot equations one can determine the physical properties immediately behind the shock (Dewey 2001, 2016). Because of its reliability, this method is often used for electronic probes calibration. This chapter deals with the comparison of shock wave time arrival measured with two piezoelectric transducers, made by different materials: a single quartz crystal and a ceramic element. The former, preferred for its low temperature sensitivity and high stability, involves expensive transducers, so representing a drawback when several sensors are required. On the other hand, a piezoelectric ceramic element, despite its high temperature-dependence and lower long-term stability, is characterized by a very high piezoelectric sensitivity and significant lower costs (Gautschi 2002). These characteristics might represent a key solution for lab-scale experimental setups where a large number of sensors are required, both in free field and confined configurations.

4.1 Experimental Setup

The experiments are carried out at ambient conditions ($T = 288\,\mathrm{K}$, $P = 10^5\,\mathrm{Pa}$), where the reactive gaseous mixture (C_3H_8/O_2) is initially confined in a hemispherical volume, realized by a soap bubble. The detonation of the stoichiometric propane-oxygen mixture is achieved by a zinc-copper wire (0.12 mm diameter) rolled up on the electrodes of the discharge system. More details are available in Chap. 6. The measurements, for blast wave characterization, are performed by several piezoelectric transducers inserted in the table, by means of a polymeric adapter, at different distances from the center of the gaseous charge. Each sensor is pre-amplified and its signal is recorded by a Dewetron DW-801. This acquisition system is able to manage up to 16 input signals with 1 MHz sampling frequency. Since years, piezoelectric transducers are largely employed in industrial and research sectors for the measurement of force, acceleration and pressure variation. The reliability of the measure is associated with the physical properties of their sensing element, which, firstly, must be characterized by a high piezoelectric sensitivity. The latter requires a high element or crystal symmetry, thus high precision in crystal cutting. Moreover, beside a high mechanical strength to sustain significant loads, a high rigidity is required to achieve high natural frequency. Considering measurements in aggressive environments, one requires a certain stability from the sensing element, in order to avoid, when subjected to high loads and temperature, twinning phenomena (Gautschi 2002). This characteristic strongly depends on the type of sensing material, thus conditioning significantly the cost of the transducer.

A crystal of quartz, accurately cut for the type of application, is able to satisfy all the technical requirements to attain a reliable measure, with a long-term stability. The main drawback is represented by its costs, which might represent a limit when a large number of transducers must be implemented in the lab-scale setup. On the other hand, the choice of piezoelectric ceramics can be a valuable solution for certain applications, due to their lower costs and compactness. They are obtained by sintering of a finely ground powdered ferroelectric mixture, whose micro crystals are then polarized by an intense electric field at a high temperature (~200 °C). Due to the material nature, the spontaneous polarization persists, when the electric field is removed, awarding the sensing element of piezoelectric properties (Gautschi 2002). Among the available ferroelectric materials, the lead-zirconite-titanate mixed ceramics (PZT) are the most commonly used. Despite the very high sensitivity achieved by ceramic elements (greater than single crystals), they suffer from few drawbacks that can limit their application as direct pressure transducers for blast characterization. In particular, when subjected to high load and temperature, their polarization can slightly decrease causing a lowering of sensitivity over time (aging). In addition, their ferroelectric properties are very sensitive to temperature, especially near the Curie value, above which the material loses its polarization (Gautschi 2002) and this strongly limits the transducer reliability in the measure of hydrostatic pressure, associated with signal voltage intensity (Dewey 2016). However, the arrival time of, at least, the primary shock can be properly captured, due to the high sensitivity of the ceramic element.

In the light of these considerations, a piezoelectric ceramic transducer (here defined as MP), provided by Metal Processing, has been compared with a Kistler 603B, made by a crystal quartz element and typically used in the above-mentioned test bench, to measure the arrival time, in two different locations (two sensors for each type), of the primary shock generated by the explosion of the reactive mixture. The Kistler is characterized by a natural frequency of about 300 kHz while that of MP is of the order of 2 MHz, with a Curie temperature of 320 °C. In Fig. 4.1, one can see the two transducers, the MP, with a diameter of 2 mm, is significantly smaller than the Kistler (about 5.5 mm diameter), resulting suitable for an easier implementation in small-scale geometrical models. For all the shots, the hemispherical volume has a radius of 40 mm, except in one case where, just for purpose of comparison, a ring of 50 mm radius is used. The former corresponds to a reactive mixture mass of 0.186 g while the second to 0.364 g. In Fig. 4.2, one can see the schematic locations of the transducers with respect to the considered configurations: free-field (F) and confined (C).

The walls of the parallelepiped confined case (see Fig. 4.2b) are made by Medium Density Fiberboard (MDF); each wall has a thickness of 40 mm and is 240 mm height. In Table 4.1, one can see the number of shots carried out and the respective

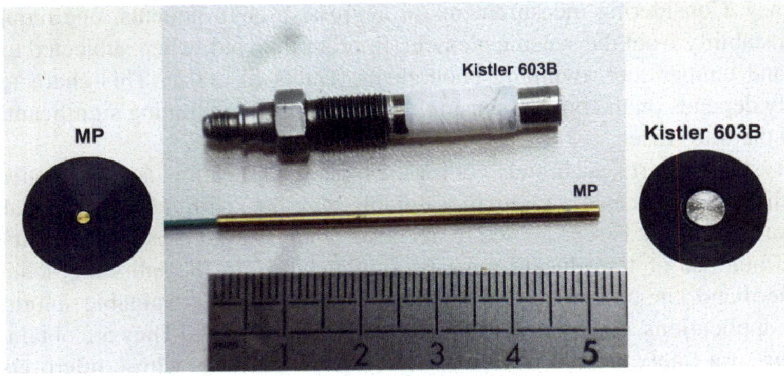

Fig. 4.1 Transducers size comparison: Kistler 603B vs MP

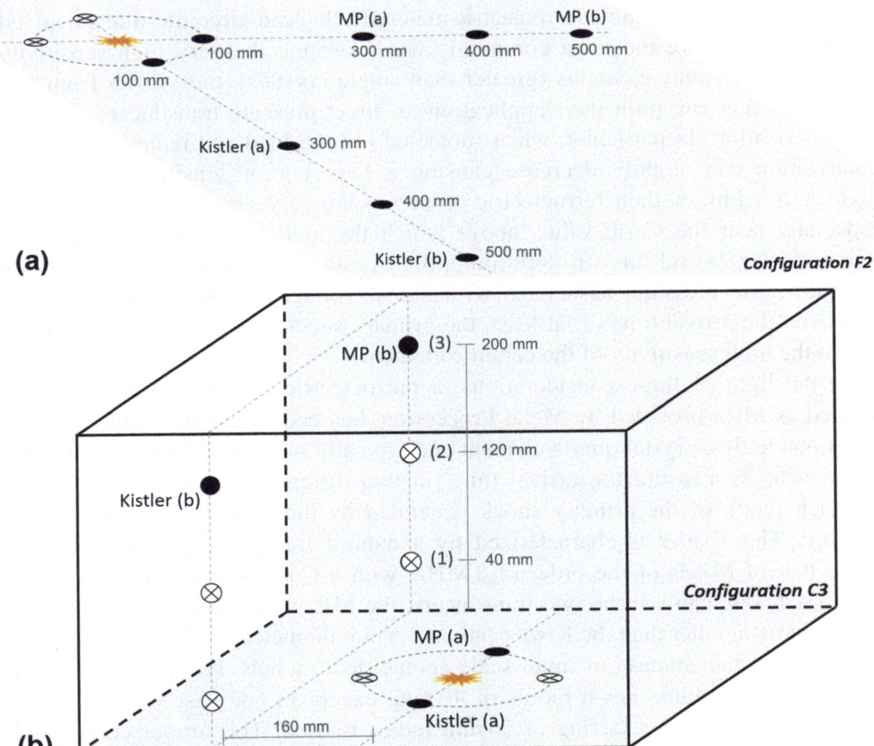

Fig. 4.2 Schematic of testing configurations and transducers position. Free field configuration $F2$ (**a**) and confined configuration $C3$ (**b**) are here reported

sensor locations. The longer wall side, where the sensors are inserted, is 600 mm long, whereas the short one is 450 mm. As presented in Fig. 4.2b, the second transducers, Kistler (b) and MP (b), are mounted in one of the available positions of the longer wall, at 160 mm from the wall vertical axis, in turn in correspondence of the hemispherical volume center.

The location, as well as the configuration reference, is denoted by a number referred to the sensor position (P), with respect to its distance from the ground ($P1 = 40$ mm, $P2 = 120$ mm, $P3 = 200$ mm).

4.2 Analysis Method

This analysis aims to verify the reliability of the MP transducer in the measurement of primary (or incident) shock arrival time, as well as in the theoretical calculation of the shock overpressure. To this end, the signal captured by Kistler gauges is assumed as baseline reference from which the relative difference between the two probes is calculated as

$$\Delta e^{\%}_{t_a} = \frac{t_a^{MP} - t_a^{KIS}}{t_a^{KIS}} \times 100 \tag{4.1}$$

where t_a^{MP} and t_a^{KIS} are, respectively, the arrival time recorded by MP and by Kistler. Of course, if the difference $\Delta t_a = t_a^{MP} - t_a^{KIS}$ is positive, the MP sensor shows a delay with respect to Kistler, while, conversely, an earlier detection if negative. Referring to Table 4.1, a mean arrival time value is calculated for each transducer, as well as the mean percentage relative difference. Moreover, by means of the standard deviation, the measures dispersion of the two different sensors is compared with respect to the different configurations.

In Fig. 4.3, one can see the voltage signal comparison recorded by the MP(a) and Kistler(a) (see Fig. 4.2) in the first shot of configuration $F1$. Before the shock passage, the signal of MP (dashed line) shows significant oscillations probably due to its lower shielding, with respect to Kistler, and higher sensitivity to electrical perturbations from the rapid discharge of the capacitor and wave propagation within

Table 4.1 Testing configurations and the respective number of shots

Sub-configuration (mm)	Sensor position (a) (mm)	Sensor position (b) (mm)	Shots number
$F1$	100	500	3
$F2$	300	500	4
$F3$	400	500	4
$C1$	100	$P1$	2
$C2$	100	$P2$	2
$C3$	100	$P3$	2

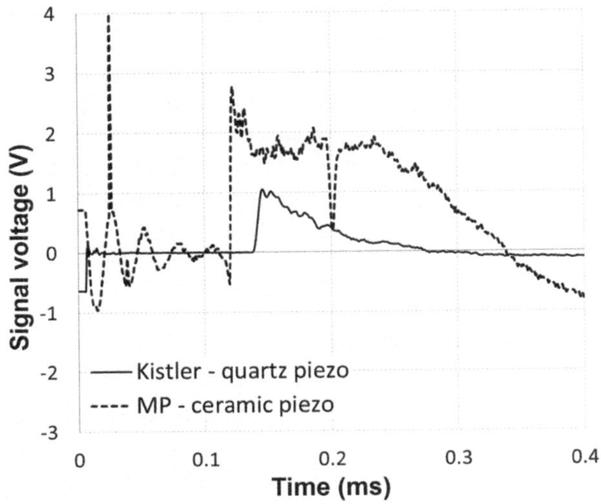

Fig. 4.3 Voltage signal recorded by MP(a) and Kistler(a) transducers. Determination of shock arrival time from incident peak by tangent method

the testing table. Moreover, the signal of Kistler quickly returns around the zero, showing a significant slower negative deviation with respect to that of MP. The arrival of the primary shock wave is identified by the peak detected, by both the sensors, around 0.08 ms, characterized by a very steep slope followed by a smooth decreasing. In general, differences of few microseconds are observed between the two transducers. In the case presented in Fig. 4.3, Δt_a is negative, with an early detection of 5 μs by MP. The arrival time value is determined by the tangent method applied to the rising slope of the shock peak.

For both the transducers, the arrival times of the primary shock recorded in configuration *F* (free field) are fitted, by least square method, using the empirical formula (Dewey 2001, 2005)

$$R = A + Ba_0t + Cln(1 + a_0t) + D\sqrt{ln(1 + a_0t)} \tag{4.2}$$

where a_0 is the sound speed of ambient air (at 20 °C) in m ms^{-1}, whereas *A*, *B*, *C*, and *D* are the fitting coefficients. Equation (4.2), which describes the shock radius as a function of time, can be differentiated to obtain the Mach number *M* change over time

$$M = \frac{1}{a_0}\frac{dR}{dt} = B + \frac{C}{1 + a_0t} + \frac{D}{2(1 + a_0t)\sqrt{ln(1 + a_0t)}} \tag{4.3}$$

To be note that, in the data fitting, the coefficient *B* is assumed constant and equal to unity, thus the shock speed will approach the ambient sound speed at asymptotically conditions (Dewey 2005). Then, for each measured arrival time one can determine the hydrostatic overpressure with the Rankine-Hugoniot equation (Dewey 2001, 2005)

$$\frac{\Delta P}{P_0} = \frac{7}{6}(M^2 - 1) \tag{4.4}$$

where P_0 is the ambient pressure. The overpressure of primary shock is calculated for all the measures of free field and for that on the ground of the confined case. The estimated results are then compared with the overpressure measurements achieved directly with the calibrated Kistler transducer.

4.3 Results and Discussion

4.3.1 Configuration F: Free Field

In Table 4.2, one can see the mean arrival time $\overline{t_a}$ of the measures obtained in the different positions of the transducers. Since for each shot a couple of sensors (MP and Kistler) was always placed at 500 mm cm from the center of the explosive charge, one mean value is calculated for all the measures in position (b) of configurations $F1$, $F2$, and $F3$. The reported errors correspond to an expanded uncertainty with a 95% confidence level.

The relative percentage difference of the mean values is calculated with respect to the arrival time of Kistler transducer, assumed as baseline reference. In the reported data, the highest discrepancy between the mean arrival times is observed for the sensors positioned at 100 mm from the charge (configuration $F1$(a)), which is the nearest locations analyzed in this work: the MP sensor provides a mean earlier detection of about 13% with respect to Kistler, whereas, in the other cases, a relative difference below 1% is achieved. The arrival times of the primary shock, for a total of 20 values, measured by the two transducer types are compared in Fig. 4.4, together with the respective fitting curves. The corresponding coefficients achieved for Eq. (4.2), reported in Table 4.3, do not reveal significant differences between the two set of data, providing a very good overlapping between MP and Kistler fitting laws, within the considered ranges.

A slight difference, toward smaller times (dashed line in Fig. 4.4), is observed in correspondence of the lowest distance from the charge (100 mm), for which a higher

Table 4.2 Mean arrival time of primary shock in free field different sensor positions: comparison between MP and Kistler (95% confidence level expanded uncertainty)

Configuration (position)	t_a^{MP} (ms)	t_a^{KIS} (ms)	Δe_{t_a} (%)
$F1$(a)	0.089 ± 0.018	0.102 ± 0.042	-13.07
$F2$ (a)	0.510 ± 0.005	0.508 ± 0.007	$+0.33$
$F3$ (a)	0.768 ± 0.012	0.767 ± 0.015	$+0.20$
$F1$ (b), $F2$ (b), $F3$ (b)	1.032 ± 0.008	1.029 ± 0.010	$+0.27$

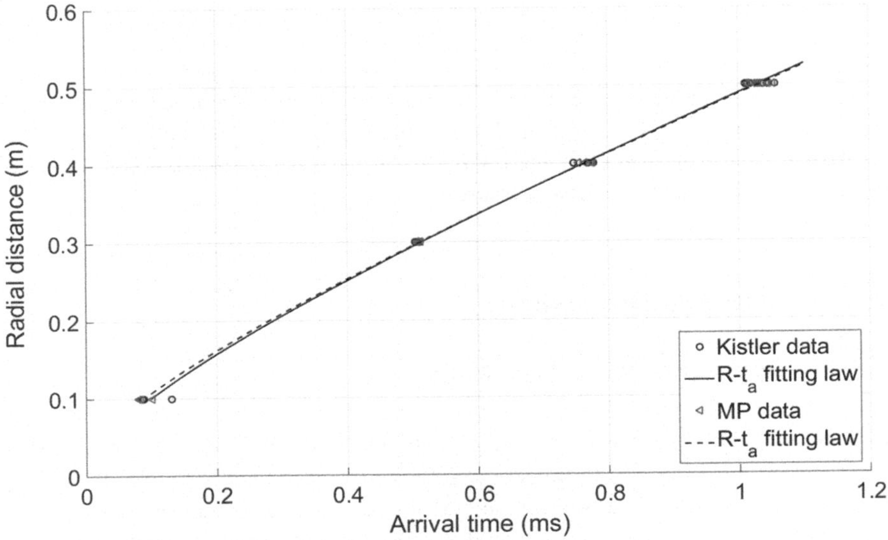

Fig. 4.4 Primary shock arrival time measures fitted with Eq. (4.2): comparison between MP and Kistler

Table 4.3 Arrival time fitting coefficients with MP and Kistler

Fitting coefficients	A	B	C	D
Kistler	−0.0025	1	−0.2754	0.4284
MP	0.0117	1	−0.2390	0.3804

relative deviation of MP measures is achieved. However, this negligible difference with regard to radius-time empirical relation (Eq. (4.2)) for the primary shock leads to consider MP as a promising cheaper alternative to Kistler transducers.

The obtained empirical coefficients of Table 4.3 are then used to determine the Mach number and, consequently, the primary shock hydrostatic overpressure for each of the measured arrival times. The mean overpressure $\overline{\Delta P}$, as a function of the distance from the gaseous charge, is presented in Fig. 4.5. The values calculated by Eq. (4.4), from MP and Kistler time measurements, are compared with the mean overpressure $\overline{\Delta P_m}$ measured by a calibrated Kistler.

In Table 4.4, the mean calculated overpressures, $\overline{\Delta P}^{MP}$ and $\overline{\Delta P}^{KIS}$, are reported with the relative percentage deviation of MP mean values with respect to Kistler. In addition, in Table 4.5, one can see the mean overpressure $\overline{\Delta P}_m^{KIS}$ and the corresponding relative deviation of calculated values for both the transducers.

In Fig. 4.5, one can see a certain agreement, which enhances with radial distance, of the mean calculated overpressure between the two transducers. At $R = 100$ mm the largest difference, of about 140 mbar, is observed, whereas in correspondence of the other sensor positions the gap remains within a maximum of 60 mbar. In particular, at all the considered radial distances, the mean MP values underestimate ΔP

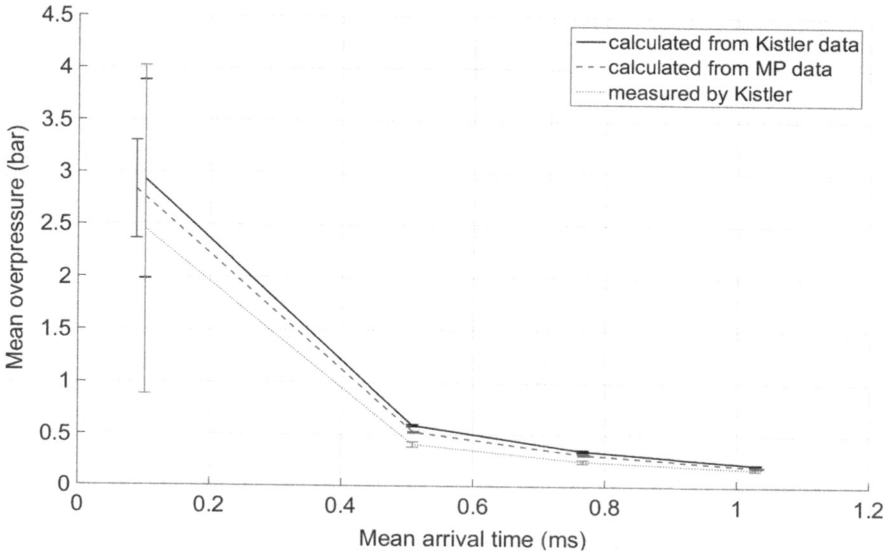

Fig. 4.5 Hydrostatic overpressure calculated by Eq. (4.4) for MP and Kistler time data in free field: comparison with Kistler pressure measures. Error bars with 95% confidence level

Table 4.4 Mean calculated hydrostatic overpressure in free field different sensor positions: comparison between MP and Kistler (95% confidence level expanded uncertainty)

Configuration (position)	$\overline{\Delta P}^{\mathrm{MP}}$ (bar)	$\overline{\Delta P}^{\mathrm{KIS}}$ (bar)	$\Delta e^{\mathrm{rel}}_{\overline{\Delta P}}$ (%)
$F1$(a)	2.83 ± 0.469	2.97 ± 0.968	−4.5
$F2$ (a)	0.52 ± 0.005	0.58 ± 0.010	−10.3
$F3$ (a)	0.31 ± 0.007	0.34 ± 0.010	−8.7
$F1$ (b), $F2$ (b), $F3$ (b)	0.20 ± 0.003	0.21 ± 0.004	−7.2

Table 4.5 Mean measured hydrostatic overpressure in free field by a calibrated Kistler transducer (95% confidence level expanded uncertainty): comparison between MP and Kistler relative deviation

Configuration (position)	$\overline{\Delta P}^{\mathrm{MP}}$ (bar)	$\Delta e^{\mathrm{rel}}_{\mathrm{MP}}$ (%)	$\Delta e^{\mathrm{rel}}_{\mathrm{KIS}}$ (%)
$F1$(a)	2.59 ± 1.178	+9.6	+14.8
$F2$ (a)	0.41 ± 0.026	+29.0	+43.8
$F3$ (a)	0.25 ± 0.017	+26.3	+38.3
$F1$ (b), $F2$ (b), $F3$ (b)	0.18 ± 0.006	+10.3	+18.9

with respect to Kistler ones, on the base of the fitting coefficients determined from the respective arrival time measurements. On the other hand, both the calculated MP and Kistler overpressures overestimate if compared to the pressure values measured by the calibrated Kistler. Also in this case, the maximum deviation is achieved close to the explosive charge, always characterized by the greatest measure dispersion both in time and pressure: an overestimation of 240 mbar for MP while of 340 mbar

Table 4.6 Mean values of the differences between MP and Kistler: measured arrival times and calculated overpressures (uncertainty with 95% confidence level)

Configuration (position)	$\overline{\Delta t_a}$ (μs)	$\Delta e_{t_a}^{rel}$ (%)	$\overline{\Delta_{\Delta P}}$ (mbar)	$\Delta e_{\Delta P}^{rel}$ (%)
$F1$(a)	13.3 ± 24.3	11.4	274 ± 95	9.3
$F2$ (a)	4.0 ± 1.0	0.8	52 ± 26	10.3
$F3$ (a)	v5.0 ± 2.2	0.7	27 ± 10	8.7
$F1$ (b), $F2$ (b), $F3$ (b)	5.0 ± 2.5	0.5	15 ± 2	7.1

for Kistler. Thus, the results achieved with MP data provide a smaller relative difference from the measured mean overpressures, especially at 100 mm from the charge, where it is below 10%. The disagreement between calculated and measured values might be associated with the hypothesis of the Rankine-Hugoniot equation (Eq. (4.4)), which assumes ideal gas and adiabatic flow (Ben-Dor et al. 2001), used to determine the overpressure from arrival time measurement. In addition, a further cause might be due to the nature of piezoelectric sensor, whose detection capability is limited to the rapid mechanical compression imposed by the shock. In fact, it has no sensitivity to the entropy decreasing radial gradient which, by affecting the movement of the particles in the flow field (Dewey 2001, 2016), is marked out by significant higher characteristic times than that of the pressure front.

In Table 4.6, one can see the mean differences in measured arrival time and calculated overpressure between MP and Kistler, as a function of the radial distance. The relative percentage of absolute difference values, with respect to Kistler, are also reported. These results, together with the corresponding uncertainty with 95% confidence, are displayed in Fig. 4.6.

Since, in this case, the attention is on the difference between the measures achieved by the two transducers, the mean difference in arrival time, as well as its uncertainty, accounts for a total of 22 time measurements, thus including one test performed with a gaseous hemispherical volume with 50 mm radius. On the contrary, the mean difference $\overline{\Delta_{\Delta P}}$ is based on 20 calculated overpressures, by means of the previously determined radius-time fitting laws (see Table 4.3).

With regard to the arrival time, as expected from the notable data dispersion, a significant difference, with a mean value of about 13 μs, is achieved at 50 mm from the center of the explosion. On the contrary, in the farther locations, the mean deviation remains stable around 5 μs, with restrained uncertainty intervals. For what concern the calculated overpressure (see Fig. 4.6 bottom), the mean difference between MP and Kistler results quite high, about 274 mbar, in the nearest charge position, followed by a gradual decrease down to 15 mbar at 50 mm radius. However, despite this, an overpressure difference of 50 mbar, as in the case of configuration $F2$(a), can be significant, leading to misleading conclusions especially with regard to safety and buildings damage analysis, for which a precise overpressure estimation is required. One must remember that the above-

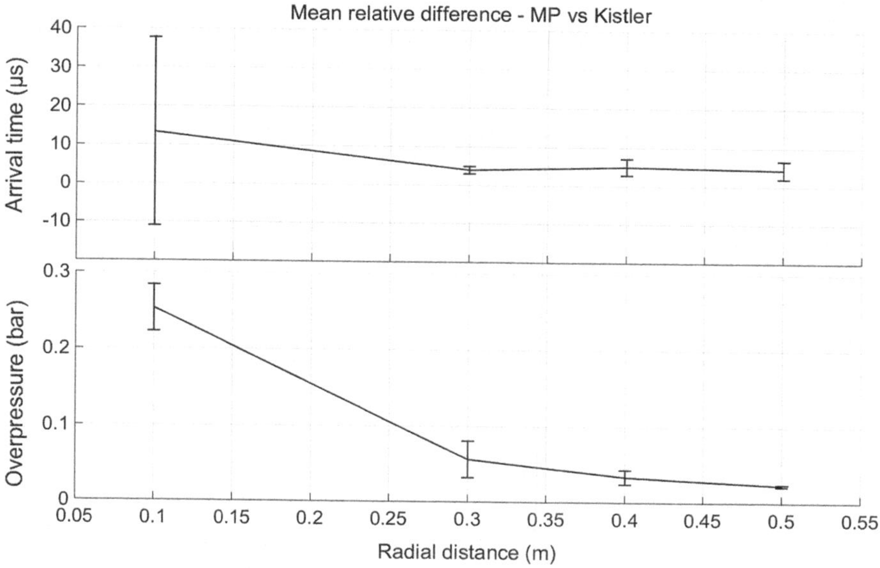

Fig. 4.6 Mean relative difference between MP and Kistler: measured arrival time and calculated overpressure with respect to radial distance. Error bars with 95% confidence level

mentioned differences refer to the calculated overpressure, whose values, for both MP and Kistler, are at least 20–30 mbar higher than that directly measures with the calibrated transducer.

Therefore, the investigated piezoelectric ceramic transducer, despite its great advantages in terms of cost, size, and sensitivity, cannot directly substitute for the use of a piezoelectric crystal sensor, which offers, by an appropriate calibration, the measure of shock overpressure with a certain level of accuracy. Indeed, due to the temperature sensitivity of ceramic material, it is not possible to use this kind of transducer to directly measure the shock overpressure, remaining limited to arrival time detection and, consequently, overpressure calculation. Nevertheless, in terms of measured arrival time and calculated overpressure, the MP ceramic sensor reveals a lower data dispersion with respect to Kistler, so providing smaller uncertainties in overpressure estimation by the Rankine-Hugoniot equation (Eq. (4.4)). In this case, the achieved values result closer to the directly measured overpressure than that calculated by the Kistler itself, but still showing a not negligible overestimation. However, the exploitation of a ceramic piezoelectric transducer should not be excluded for all the application since, for instance, its higher repeatability in time measurements might lead to prefer it for radius-time empirical law determination, when a notable number of sensors are required in medium scale test benches.

Table 4.7 Mean arrival time and mean calculated hydrostatic overpressure of primary shock in configuration C(a): comparison between MP and Kistler (95% confidence level expanded uncertainty)

Configuration C(a)	MP	Kistler	Δe^{rel}(%)
$\overline{\Delta t_{\mathrm{a}}}$ (ms)	0.090 ± 0.006	0.096 ± 0.008	−6.3
$\overline{\Delta P}$ (bar)	2.80 ± 0.15	3.03 ± 0.20	−7.7

Table 4.8 Mean measured hydrostatic overpressure in confined geometry by a calibrated Kistler transducer (95% confidence level expanded uncertainty): comparison between MP and Kistler relative deviation

Configuration (position)	$\overline{\Delta P_m^{\mathrm{KIS}}}$ (bar)	$\Delta e_{\mathrm{MP}}^{\mathrm{rel}}$ (%)	$\Delta e_{\mathrm{KIS}}^{\mathrm{rel}}$ (%)
C (a)	2.57 ± 0.11	+9.0	+18.1

4.3.2 *Configuration C: Confined Geometry*

In confined geometry the analysis is limited to the measurements obtained on the ground, at a radial distance of 100 mm from the charge center (see Fig. 4.2). As previously discussed, at this location one observes the highest measure dispersion, for both MP and Kistler sensors. Two shots are carried out for each of the three configurations, leading to the collection of six time values, in turn used to calculate the shock overpressure by means of Eq. (4.4). To this end, the corresponding Mach number is estimated with the fitting coefficients determined in the free field case (see Table 4.3).

The mean measured arrival time and the mean calculated overpressure are reported in Table 4.7. The arrival time deviation of MP, with respect to Kistler, is still negative, but around a half of the corresponding value in configuration $F1$(a). On the contrary, the relative difference in the calculated mean overpressure is increased up to 7.7%. This is due to a little higher $\overline{\Delta P}$ value (+2%) calculated from Kistler time data, whereas that calculated from MP, remaining closer to its correspondent value at $R = 100$ mm in free field, is decreased of 1.1%. In Table 4.8, one can see the mean overpressure $\overline{\Delta P_m^{\mathrm{KIS}}}$ and the corresponding relative deviation of calculated values for both the transducers. Moreover, in Fig. 4.7, the mean overpressure values, calculated from MP and Kistler data, are compared with that measured by the calibrated Kistler.

With regard to the deviation from the measured overpressure, both the calculated mean values provide an overestimation, but the overpressure of MP still results closer to the measured one, with about 9% difference. Therefore, as discussed in Sect. 4.3.1, the significant overestimation in the calculated overpressure does not allow to rely on the considered piezoelectric ceramic sensor to determine the shock pressure, especially in confined geometries, where the accuracy in overpressure measure is a key element.

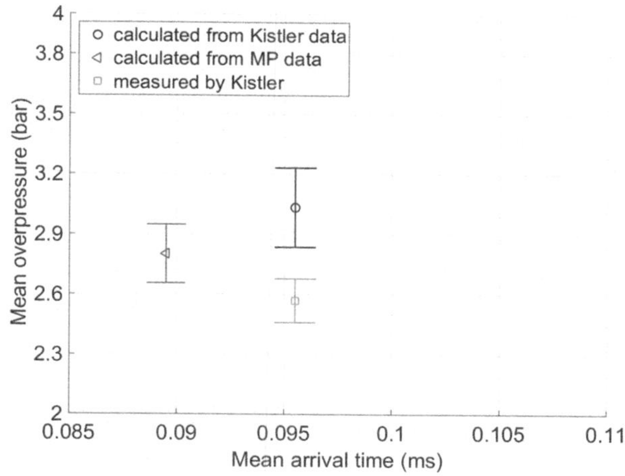

Fig. 4.7 Hydrostatic overpressure calculated by Eq. (4.4) for MP and Kistler time data (ground transducers) in confined geometry: comparison with Kistler pressure measures. Error bars referred to the expanded uncertainty with 95% confidence level

For what concern the measures obtained by the sensors placed on the wall, only qualitative considerations are proposed, since the analysis approach described in Sect. 4.2 can be only applied to the primary shock. Indeed, the arrival time detected on the wall is associated with a reflected shock wave. In Fig. 4.8, one can see the comparison between MP and Kistler voltage signals in the three different wall locations, from the bottom ($P1$) to the top one ($P3$). A quite good agreement seems revealed in the detection of the first reflected wave corresponding peak.

Nevertheless, a significant different behavior can be observed for MP and Kistler transducers. The latter demonstrates a more stable signal, by quickly returning around zero after a pressure peak and showing a limited negative signal deviation. The former, on the contrary, is characterized by a slow signal post-peak decrease together with notable deviations in the negative region. Such behavior might be reasonably correlated with the temperature sensitivity of the ceramic element, which involves the observed signal offsets, which does not stabilize anymore around zero after the detection of the first peak. In fact, in Fig. 4.8, one can see the detection of further reflected wave peaks, whose rising points are affected by large voltage offset. At least three peaks can be clearly watched, for both the transducers, in configurations $C1$ and $C2$, with a possible, less evident, fourth one. In configuration $C3$, the farthest from the gaseous reactive charge, only two peaks seem well visible, while the third and the fourth, despite effectively observables in MP signal, appear very weak in Kistler one.

Finally, in Fig. 4.9, the difference in arrival time between MP and Kistler is compared with respect to the detected peaks. The displayed curve values refer to the average between the two shots realized for configuration $C1$. From this comparison, any particular behavior is revealed, except for a possible growing trend of the gap between the time detected, respectively, by the two sensors, as observed for configuration $C2$ and $C3$, up to the third peak.

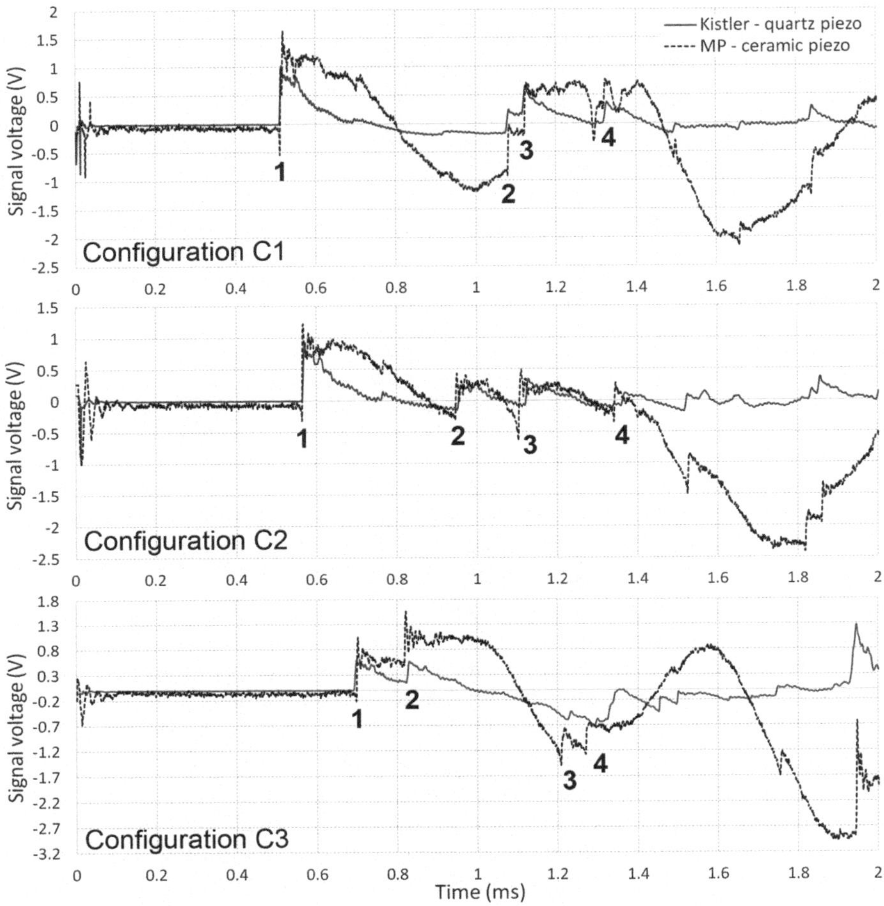

Fig. 4.8 Voltage signal comparison between MP and Kistler of test 1 in confined geometry for wall transducers in position *P*1, *P*2 and *P*3

In conclusion, in confined geometry, even if its signal appears strongly affected by temperature, the piezoelectric ceramic transducer seems able to detect multiple reflected wave peaks. Nevertheless, except for the first reflected wave, a certain difference is attained with respect to the arrival times recorded by Kistler. However, stating the previously argued considerations about the overpressure calculation in relation with the limits of the ceramic material, further investigations might be taken into account. In particular, as one can see in Fig. 4.8, the voltage intensity of the first peak in MP signal presents a reasonable decrease as the sensor moves farther from the explosive charge (from *P*1 to *P*3). Thus, it might be eventually possible to calibrate the MP sensor by means of Kistler gauge, carrying out an analysis focused on the measure of only the first reflected wave pressure. In such a case, a larger

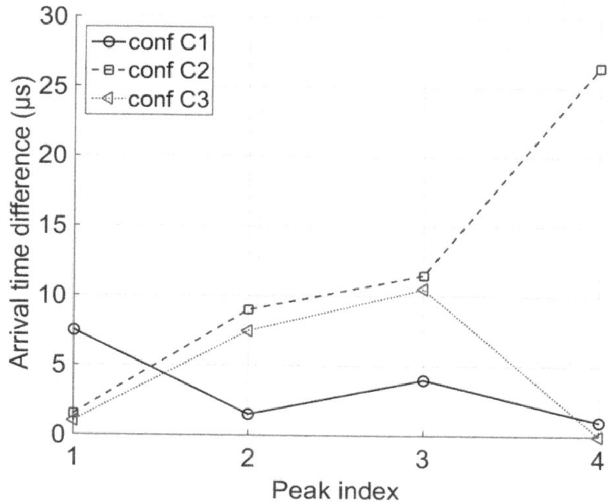

Fig. 4.9 Mean arrival time difference between MP and Kistler sensors on reflected wave peaks

number of shots would be required, in order to evaluate the MP sensor in several locations of the confined geometry, as well as to highlight possible aging of the ceramic element besides, of course, the effects of temperature.

References

Ben-Dor, G., Igra, D., & Elperin, T. (2001). *Handbook of shock waves* (Chap. 2, Vol. 1). London: Academic.

Chue, S. H. (1975). Pressure probes for fluid measurement. *Progress in Aerospace Sciences, 16*, 147–223.

Dewey, J. M. (2001). Spherical shock waves: Expanding spherical shocks (blast waves). In G. Ben-Dor, D. Igra, & T. Elperin (Eds.), *Handbook of shock waves* (Chap. 13, Vol. 2). London: Academic.

Dewey, J. M. (2005). The TNT equivalence of an optimum propane-oxygen mixture. *Journal of Physics D: Applied Physics, 38*, 4245–4251.

Dewey, J. M. (2016). Measurements of the physical properties of blast waves. In O. Igra & F. Seiler (Eds.), *Experimental methods of shock waves research*. Cham: Springer International Publishing.

Gautschi, G. (2002). Piezoelectric materials for sensors. In *Piezoelectric sensorics*. Berlin: Springer.

Josey, T., Ritzel, D. V., & Sawyer, T. W. (2016). Development of a miniature double Pitot-static probe and its application to calibrating blat flow conditions. In *24th Int. Symp. on Military Aspect of Blast and Shock, Halifax, Canada.*

Julien, B., Sochet, I., & Vaillant, T. (2016). Impact of the volume of rooms on shock wave propagation within a multi-chamber system. *Shock Waves, 26*, 87–108.

Kinney, G. F., & Graham, K. J. (1985). *Explosive shocks in air*. Berlin: Springer.

Miura, A., Matsuo, A., Mizukaki, T., Shiraishi, T., Utsunomiya, G., Takayama, K., et al. (2004). Reflection and diffraction phenomena of blast wave propagation in nuclear fuel cycle facility. *JSME International Journal Series B Fluids and Thermal Engineering, 47*, 287–292.

Reisler, R. E., Keefer, J. H., & Ethridge, N. H. (1995). MABS Monograph, Air Blast Instrumentation 1943–1993: Measurement Techniques and Instrumentation. Defense Nuclear Agency.

Remennikov, A. M. (2003). A review of methods for predicting bomb blast effects on buildings. *Journal of Battlefield Technology, 6*, 5–10.

Schelinski-Glück, G. (1993). Blast propagation in tunnels behind chambers from cylindrical H.E-Charges detonation in the tunnel entrance. In *13th Int. Symp. on Military Aspects of Blast and Simulation, The Hague, The Netherlands*.

Settles, G. S. (2012). *Schlieren and shadowgraph techniques: Visualizing phenomena in transparent media*. Berlin: Springer Science & Business Media.

U.S. Department of the Army. (1990). Structures to resist the effects of accidental explosions. Technical Manual 5-1300.

Zel'dovich, Ya. B., & Raizer, Yu. P. (1966). *Physics of shock waves and high-temperature hydrodynamic phenomena* (Vol. 1). New York: Scripta Technica.

Zel'dovich, Ya. B., & Raizer, Yu. P. (1967). *Physics of shock waves and high-temperature hydrodynamic phenomena* (Vol. 2). New York: Scripta Technica.

Chapter 5
Exploding Wires

Xavier Rocourt and Isabelle Sochet

5.1 Description of Exploding Wire Phenomenon

The first reported experiments dealing with exploding wire occurred in the eighteenth century (Hansen 2011; McGrath 1966). In 1774, an English instrument maker, Nairne, observed the exploding wire phenomenon for the first time. Then two centuries passed before scientists really investigate this field of research and find practical applications for exploding wires. Despite the renewed interest of exploding wires since works of Anderson in 1920s (Anderson 1922), several aspects of the phenomenon are still not clarified. Later, the main technic used to understand the phenomenon was the visualization, performed with high speed cameras (Kerr cell, streak or rotating mirror, etc.). In 1960, Turner observed with a Kerr cell camera the location of the shock front and of the contact surface.

Chace (1959) has investigated exploding wires phenomenon with a high-speed photographs (0.3 μs) in 1959 and observed that during the early stage of a wire explosion there is a condition of extreme super-heating of the liquid. When this condition is reached, gas nuclei appear and create a discontinuous structure made of a foam metal bubbles dispersed inside the liquid followed by sudden explosive vaporization called "transplosion", a phase transformation of explosive violence. In 1999, Pikuz et al. observed a similar foam-like structure using the technique of pulsed point-projection radiography of the exploding wire.

In 2002, Taylor investigated processes that occur during the condensed wire fragments formation and the establishment around them. This study provides an explanation of the exploding wires phenomenon thanks to a fast framing digital camera (1–10 μs interframe times with about a 0.1 μs exposure) and a flash X-ray

---

X. Rocourt (✉) • I. Sochet
Laboratoire PRISME, INSA Centre Val de Loire, Bourges, France
e-mail: xavier.rocourt@insa-cvl.fr

I. Sochet (ed.), *Blast Effects*, Shock Wave and High Pressure Phenomena,
https://doi.org/10.1007/978-3-319-70831-7_5

system. The tests were performed with a 1 mm diameter copper wire. The charge voltage used was about 10 kV, the delivered energy ranged between 1.5 and 30 kJ and the discharge times were fixed between 0.8 and 10 ms. Taylor (2002) noticed on fast framing camera images that plasma spots appear and develop during the condensed wire fragments formation, just prior to voltage peak. The development occurs in the radial direction with three consecutive steps: First, a sudden expansion to between 5 and 12 mm, then a pause and finally a slow expansion to many tens of millimeters. It has been observed that the rate of expansion increased with the power discharge. Indeed, the initial rate of expansion was greater than 1200 m s^{-1} and the lower rate of expansion reached a value of 400 m s^{-1} for higher powered discharges (250 MW). On the contrary, it was 14 m s^{-1} followed with an important pause for very low power discharges (0.4 MW).

The development of the breaks, due to hot spot formation, involves that the electrical resistance increases up to a certain value where an arc discharge occurs. The condensed parts of the wire are in liquid phase, and the first electric arcs appear just after the melting period. The vapor generated by the breaks expands and the resulting plasma from the arc discharge undergoes an initial fast expansion, then continues at a slower expansion rate. An energy balance exists between the Ohmic heating and the heat loss. Initially, the expansion is great because the Ohmic heating is much greater than heat loss as the arc is small. Once the energy balance is near equilibrium, the radial expansion slows or often stops. After what the current is forced to leave and re-enter the condensed fragments of copper to bridge the break. Taylor (2002) suggested that a poorly conducting vapor boundary layer (thickness less than a few tens of microns) will exist between the condensed surface and the plasma. In the proposed model, the vapor expands away from the break due to pressure or momentum and the electrical energy heats the vapor up to temperatures typical of plasma. Energy is then transferred to the condensed material surface by both radiation and conduction, which increases the temperature to the boiling point. The appearance of striations could be caused by a nonhomogeneous heating. Next, the energy distribution along the wire becomes more uniform as the wire is completely enclosed by conducting plasma which can be assimilated to a discharge channel. The radial expansion of the wire has been found to be the dominating phenomena of the wire explosion in free air.

Recent research by Romanova et al. (2015), shows differences in the structures of these discharge channels during nanosecond discharges in fine wires made of three classes of materials: copper, tungsten, and nickel-type groups. These groups are different by their properties like the boiling temperature, heat of evaporation, electric conductivity, etc. The two extremes groups are copper-type group and tungsten-type group, the nickel-type group being intermediate. Experiments were performed using a charging voltage of 20 kV maximum, with a 10 kA peak current and a current rise rate of 50 A ns^{-1}. Observations were made with shadow photography method using a three-channel laser system. Wires were 25 μm in diameter and 12 mm long. Romanova et al. (2015) observed differences between copper and tungsten wires after the beginning of rapid expansion. Indeed, the radial expansion of the wire occurred earlier for copper (40–50 ns) than for the tungsten (80–90 ns). The diameter of the expanding wire core was larger for metals of the first group (copper, silver, and aluminum) as compared to metals of the second group (tungsten, molybdenum, and titanium). It can be explained by the longer time of energy deposition in the wire, due to the shunting discharge which occurred sooner in metals of the first group. Romanova et al. (2015) also observed the front of the shock wave generated by the exploding wire. The measured shock wave velocity was ranging from 2000 to 5000 m s^{-1}, depending on the wire material. Metals of the second group generated higher shock waves velocities.

Romanova et al. (2015) noticed that for metals of the second group, the core was surrounding by a coat which is non-conducting. For nickel-type metals (third group), it was observed that the inner core coexisted with the coat, occupying a large volume. It would appear that the coat is the part of the core that had no time to develop in full measure (Romanova et al. 2015). Moreover, both, the coat and the inner core are no conducting. For metals of the first group (copper-type) which have a good conductivity, only the coat exists.

Sarkisov et al. (2004) studied the state of the wire core in vacuum according to the amount of energy deposited before the voltage breakdown and the heating process. A 60 kV high-voltage supply was used with an energy of 12.6 J and a current rate varying from 22 and 150 A ns^{-1}. Several metallic wires (Ag, Al, Cu, Au, Fe etc.) were tested with diameters ranging from 4 to 38 μm. It has been observed that the wire remained unaltered when the deposited energy before voltage breakdown was less than the solid-state enthalpy. If the deposit energy was between the solid-state enthalpy and the liquid-state enthalpy, i.e. during melting, the wire disintegrates into macropieces and the expansion velocity was about 100 m s^{-1}. If the deposited energy was between the liquid-state enthalpy and the vaporization energy, i.e. before vaporization, the wire disintegrates into hot liquid clusters of submicron size and the expansion velocity reached 1000 m s^{-1}. Finally, for energy higher than the vaporization energy, the wire core exists in a gas-plasma state and the expansion velocity was between 2000 and 6000 m s^{-1}.

The study of the first current pulse happening during the exploding wire process is also of interest since it determines all the subsequent behavior of the phenomenon. In particular, Nash and Olsen (1964) investigated one of the most important aspects of the first current pulse, the time to burst (*Tb*) for wire explosion in air. Experimental device consisted in a condenser bank with voltages between 10 to 20 kV and energies between 1.42 to 11.4 kJ. Wires made of different materials of about 70, 150 and 180 mm long were tested. Diameters were ranging from 0.08 to 1.29 mm. Experimental results highlighted that for all the considered materials, the time to burst decreases with the wire cross section, the wire long and when the initial voltage increases. Concerning the influence of the material, for a fixed wire size and initial voltage, the order of increasing time to burst was: $Pb < Al < Au < Ag < Cu \sim Pd$.

Bennett (1958) investigated the shock wave generated by exploding wires. A 28 kV high voltage supply was employed with an energy of 118 J. A rotating mirror camera allowed to make the shock wave visible, with copper wires 20 mm long and 0.127 mm in diameter. Bennett (1958) visualized the boundary of the main glow and concluded that it corresponds to the contact surface separating the hot gases of the explosion from the surrounding atmosphere. Both theoretical and experimental works showed evidence of the presence of an inward propagation of a compression in the main shock wave in cylindrical explosions. This phenomenon was already noticed by a theoretical and experimental approach for spherical blast (Schardin 1954), with the propagation of a second shock wave in burnt gases. Finally, the study of Bennett (1958) confirmed that the behavior of the main shock wave from an exploding wire was similar to a strong cylindrical blast wave over a considerable portion of its trajectory.

An important study has been carried out by Oshima (1962) on strong cylindrical blast waves produced during the explosion of wires in a shock chamber at various low pressures, using interferometric measurements to observe density distributions of air. Oshima (1962) proposed a new theoretical model, called the quasi-similarity theory, which is based on the assumption that the distributions of the density, pressure, and flow velocity can be assumed to be locally similar. Experiments were performed with copper wires of 0.1, 0.12 and 0.15 mm diameter, using voltages between 4 and 7 kV and energies in the range of 32 to 196 J. Oshima (1962) suggested that a rarefaction wave is generated due to the sudden interruption of the core expansion. This wave is formed in the early stages of the explosion within the main wave. If the rarefaction wave catches up with the shock front, observed when Mach number is about 1.15, a secondary shock wave appears.

The shock wave formation in air was also investigated by Volkov et al. (1967) during the initial stage of wire explosion by the mean of shadow photography. A wide range of wires samples of different metals ranging from 0.03 to 0.8 mm in diameter were studied. A 30 kV high-voltage supply was used with a maximal energy of 2.7 kJ. Volkov et al. (1967) noticed, for wire diameters of around 0.2 mm, a first shock wave of about 1000 to 2000 $\mathrm{m\,s^{-1}}$ at the beginning of the outward diffusion of the metal vapors, followed by a second shock wave of 1000 to 5000 $\mathrm{m\,s^{-1}}$ appearing with the start of the dwell phase (the current flow ceases for a while). It was observed that the propagation rates of the first and second shock wave increased with the condenser voltage using 0.17 and 0.38 mm wire diameters. Moreover, the time interval between first and second shock waves decreased with the wire diameter and when the condenser voltage increased. Among speculations made about the second shock wave, Volkov et al. (1967) suggested that the most probable could be the speed of the metal vapor expansion which increases due to the rapid collapse of the magnetic field as the current dwell phase begins.

Brossard (1970) investigated the spherical shock wave propagation in air produced by exploding wires. A U shaped platinum wire was positioned on the center of a spherical bomb in order to generate a quasi-spherical shock wave. The wire was 15 mm long, 0.1 mm in diameter and was exploded by means of the discharge of a capacitor having a voltage in the range of 3–4 kV with a delivered energy of 150 and 300 J. The quasi-sphericity of the first shock wave development was examined by Doppler effects and schlieren photography. The second shock wave was also visualized, with an initial propagation in the same direction as the first shock, before going back and reflecting after focalization at the symmetry center. The intensity of the second shock wave was weak with a not perfect symmetry. Sometimes, two secondary shock waves, a few centimeters distant, were reflecting after focalization at the center of the sphere.

The variation of the thermodynamic parameters just behind the shock wave and the exact amount of energy transferred to gaseous medium surrounding the generator used to create the shock wave were characterized.

Several studies were performed about exploding wires, with the goal to initiate a spherical detonation in a reactive gaseous mixture.

In 1953, Manson and Ferrié concluded that the intensity of the shock wave generated by the ignition source is a primordial but not a sufficient factor to ignite a spherical detonation in a gas mixture. Indeed, properties of the mixture in terms of chemical reactivity are of importance since the reaction needs to be rapid enough and provide enough amount of energy. Litchfield et al. (1963) conducted experiments about exploding wire discharges to generate a direct initiation of detonation in gaseous mixtures inside two spherical bombs of 1 *L* and 14 *L*. Three mixtures of propane, ethylene, hydrogen mixed with oxygen were tested for several equivalence ratio. The minimum stored energy in the capacitor, required to ensure a direct detonation was greater with a spark ignition as compared to exploding wire ignition. Indeed, a ratio of 6.5 was found between the minimum stored energy with spark and with exploding wire, at the minimum initiation-energy composition in hydrogen-oxygen mixtures. This ratio value depended on several parameters such as the type of gas, the equivalence ratio, and the initial pressure. It was assumed that the larger efficiency of the exploding wire was associated with a greater efficiency of energy transfer to the gas.

Later, in 1965, Lee mentioned in his work the existence of a critical energy associated with the finite reaction zone thickness of the mixture. This critical blast energy required for direct initiation of detonation is proportional to the reaction zone thickness which depends on the experimental parameters (mixture composition, initial pressure).

The main conclusion of Brossard study (Brossard 1970) was that the nominal energy released in the electric circuit of the exploding wire was not the only parameter to take into consideration to propagate a spherical detonation in a gaseous reactive mixture. It is also fundamental to know the evolution in time of the pressure behind the shock and to compute the duration of its application to define the thickness of the shock wave that plays a fundamental role in the initial phase of the detonation propagation inside a reactive medium.

Then, Knystautas and Lee (1976) showed that two parameters are essential to initiate a direct detonation, with spark ignition. These parameters are the peak power of the source and the energy release until the peak power is achieved. The minimum power should permit to the source to generate a shock wave of a certain minimum strength which corresponds to the auto-ignition limit of the explosive mixture. Concerning the minimum energy, it should enable that the shock wave be maintained at or above this minimum strength for a certain duration, which corresponds to a certain minimum distance. This minimum distance, associated with a critical radius, according to the "kernel" theory (Lee and Ramamurthi 1976), is equal to the detonation cell length of the explosive mixture.

In order to complete this literature survey, next sections of this paper deal with an experimental parametric study performed in the PRISME laboratory. The influence of the wire length, wire diameter and charged voltage were investigated on the pressure profile generated by an exploding wire. Then, the transmission of the shock wave to ignite a direct detonation in a combustible gaseous mixture was observed.

5.2 Experimental Setup

Experiments realized at PRISME laboratory (Campanelli 2016) were performed on a workbench. The wire to be studied was connected between two rods at a height of 8 mm from the bench test for all tests except for the wire length test where it is 18 mm. The separating distance of the rods was adjusted between 5 and 30 mm. In order to explode the wire, a high voltage supply was connected to the rods. The total capacitance of 8 μF was charged at the desired voltage ranging from 4 to 10 kV. The peak current reached about 13 kA.

A piezoelectric high frequency pressure sensor PCB 113B21 located from 98 mm to the ignition has been employed to measure the pressure generated by the exploding wire in free air. Concerning the ignition of detonation study in propane-oxygen mixture, this pressure sensor was replaced with a PCB 113B26 which allows to measure higher overpressures. The initial confinement was delimited with a thin metal ring with a radius of $R = 47$ mm.

5.3 Experiments with Thin Copper Wires

This study focuses on the explosions of copper wires with different diameters d, varying lengths l and various capacitor charging voltages U.

The aim of these experiments is to understand the effects of those parameters on the pressure profile of the shock waves generated by the exploding wire, in terms of both peak overpressure ΔP_{max} and rise time Δt_{pmax} (Fig. 5.1).

These tests were performed in free air, then in a stoichiometric propane-oxygen mixture to observe the influence of each parameters on the onset of a detonation. One criteria characterizing a detonation is the rise time Δt_{pmax} which is less than 10 μs. The difference between the overpressures caused by an exploding wire in free air and within an oxygen-propane gaseous bubble is around one order of magnitude. When detonation occurs (Fig. 5.2), the maximal pressure measured at 98 mm from the ignition is near or over 3 bar (function of experimental conditions: length, diameter, voltage) whereas it is less than 0.7 bar in case of deflagration.

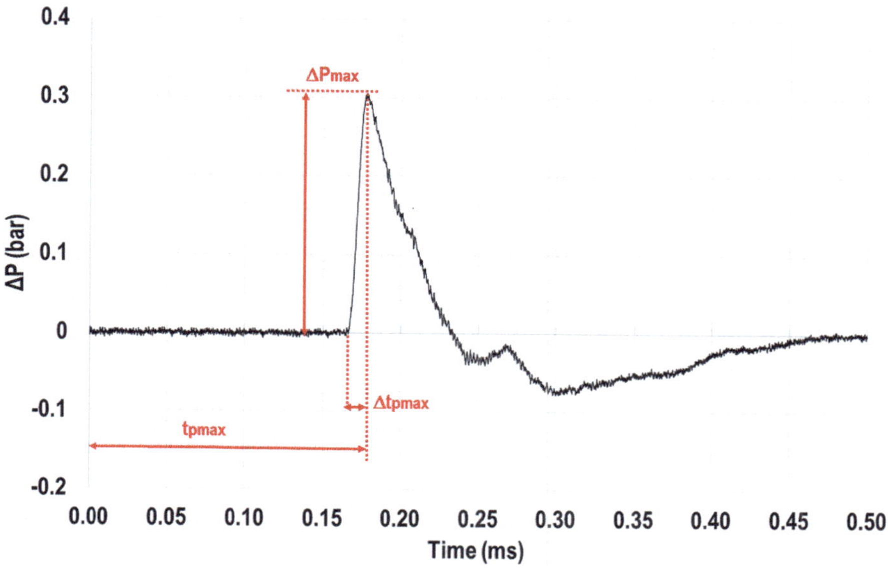

Fig. 5.1 Typical peak overpressure profile from 98 mm to a copper wire explosions in free air ($U = 8\,\text{kV}$; $d = 0.23\,\text{mm}$; $l = 6\,\text{mm}$)

Moreover, the peak overpressure is reached much earlier for the gaseous detonation, at about $\Delta t_{\text{pmax}} = 0.085\,\text{ms}$ after the trigger onset against an average value of around $\Delta t_{\text{pmax}} = 0.2\,\text{ms}$ for a deflagration.

All results presented are an average of three shots or more. The overpressure values are given with an accuracy less than 3% for exploding wire tests in air and less than 5% for gaseous explosion tests. Values of rise times are presented with an uncertainty of $\pm 0.2\,\mu\text{s}$.

5.3.1 *Influence of the Wire Diameter*

Six different copper wire diameters were studied: 0.10, 0.17, 0.23, 0.27, 0.32, and 0.37 mm. The separating distance of the rods was fixed to 6 mm for all these tests. The capacitor charging voltage was kept constant to a value of 8 kV. The nominal energy delivered through the wire was $E = \frac{1}{2}C.U^2 = 256\,\text{J}$, with C the capacity (Farad) and U the charging voltage (Volt). Results in terms of maximal overpressure and rise time according to the wire diameter in free air are reported in Fig. 5.3.

It can be observed that the resulting trend is that the peak overpressure linearly increases up to a maximum value ($\Delta P_{\text{max}} = 0.304\,\text{bar}$) obtained for a diameter of 0.23 mm and decreases thereafter. On the other hand, the rise time has a slight

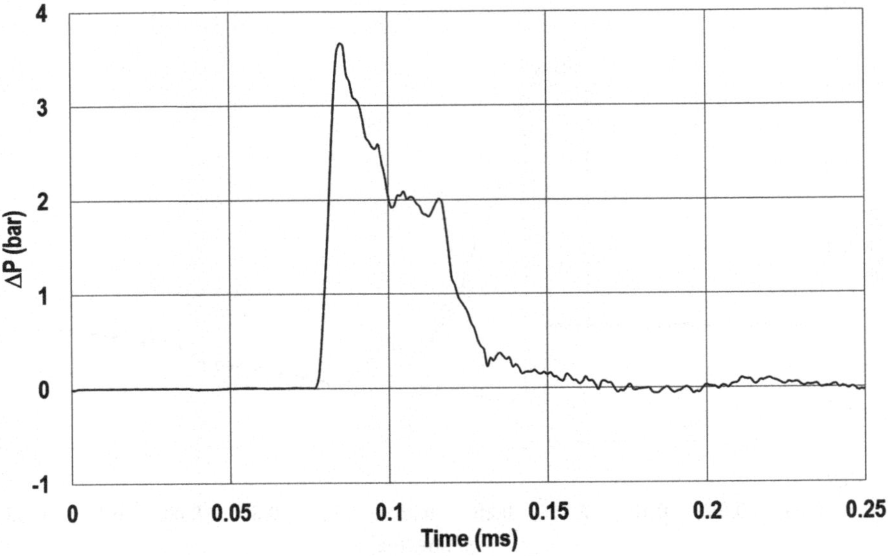

Fig. 5.2 Typical peak overpressure detonation profile from 98 mm to a copper wire explosion in a $C_3H_8 - 5\ O_2$ mixture ($U = 8\,\mathrm{kV}$; $d = 0.23\,\mathrm{mm}$; $l = 6\,\mathrm{mm}$; $R = 47\,\mathrm{mm}$)

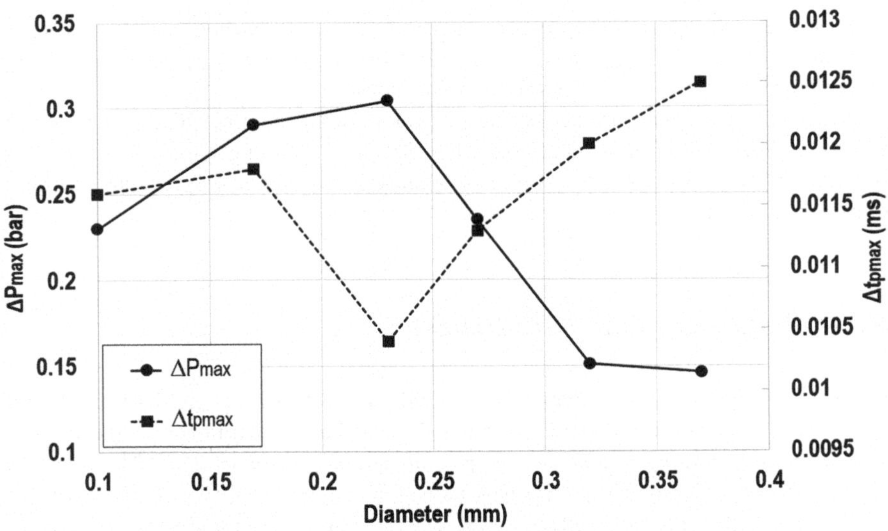

Fig. 5.3 Maximal overpressure and rise time according to the wire diameter d from 98 mm to a copper wire explosion ($U = 8\,\mathrm{kV}$, $l = 6\,\mathrm{mm}$)

Table 5.1 Maximal overpressure measured at 98 mm from the ignition—Diameter effect

$C_3H_8 + 5\,O_2 - R = 47\,\text{mm} - U = 8\,\text{kV} - l = 6\,\text{mm}$			
Diameter (mm)	$\Delta P_{max}/\Delta t_{pmax}$ wire (bar ms^{-1})	ΔP_{max} deflagration (bar)	ΔP_{max} detonation (bar)
0.10	19.82	0.409	–
0.17	24.58	–	2.908
0.23	29.42	–	3.259
0.27	20.74	0.668	2.720
0.32	12.58	0.569	–
0.37	11.68	0.468	–

increment with the thickness of the wire, with the exception of the 0.23 and 0.27 mm diameter. The diameter of 0.23 mm generates a pressure wave with the smallest rise time ($\Delta t_{pmax} = 0.0104$ ms).

In a comparable manner to the work done for the explosions of wires in free air, different diameters were used to ignite the propane-oxygen mixture, while keeping constant both the voltage ($U = 8$ kV) and the wire length (6 mm). Detonations were obtained only for three diameters: 0.17, 0.23, and 0.27 mm. This is in line with what has been found out in the earlier investigations, where the maximum overpressures and low rise times were obtained for these three diameters. Results are reported in Table 5.1 and in Fig. 5.4 where ratios of the peak overpressure and the rise time are plotted against the wire diameter. For the case of $d = 0.27$ mm there is a condition where both deflagrations and detonations have been observed with a ratio of about 50%. It seems that for a given combustible mixture, a critical value exists for the ratio $\Delta P_{max}/\Delta t_{pmax}$ generated by an exploding wire to ignite systematically a detonation. In these experiments, this value is between 20.7 and 24.6 bar ms^{-1} for a stoichiometric propane-oxygen mixture, at a distance of 98 mm from the ignition source.

The generated overpressure observed for the detonation case increases with this ratio value up to 3.259 bar measured at 98 mm from the ignition. Indeed, the distance of transition to detonation should decrease with the ratio $\Delta P_{max}/\Delta t_{pmax}$ value up to obtain a direct detonation and to generate the maximal overpressure.

Consequently, the wire diameter should be large enough to deliver a sufficient amount of energy in a short time, but if it is too high, the time to burst is too important (Nash and Olsen 1964) to initiate a detonation.

5.3.2 *Influence of the Wire Length*

Explosions were initiated by means of movable electrodes that enabled to use wire lengths ranging from 5 mm up to 30 mm. The height of the wire was maintained at 18 mm from the test bench and the diameter was kept constant ($d = 0.23$ mm). The capacitor charging voltage was fixed to 8 kV, which corresponds to a nominal energy

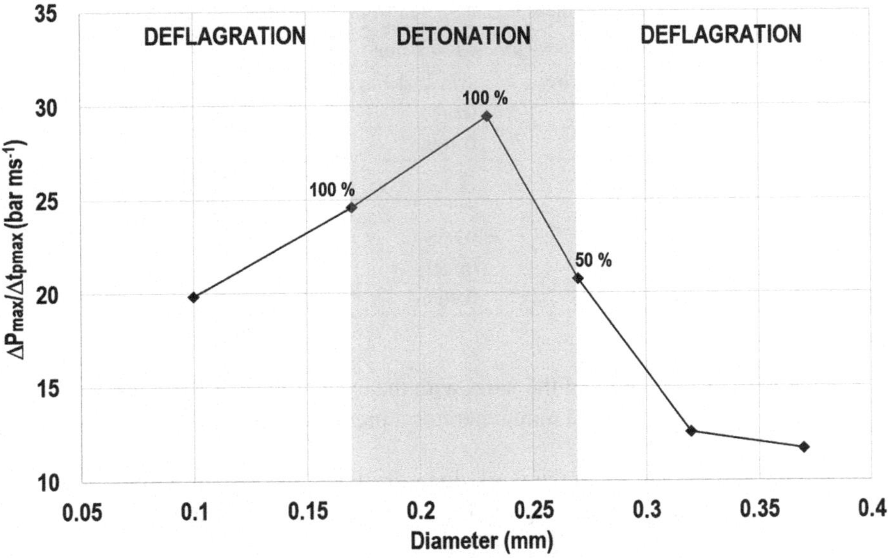

Fig. 5.4 Maximal overpressure-rise time ratio according to the wire diameter d. Percent represents the probability to obtain a detonation ($U = 8\,\text{kV}$, $l = 6\,\text{mm}$, $R = 47\,\text{mm}$)

delivered through the wire of 256 J. Experimental values in free air of maximal overpressure and rise time according to the wire length are reported in Fig. 5.5.

The experimental results suggest that a maximum value of the overpressure is reached for a length of 15 mm ($\Delta P = 0.518\,\text{bar}$ at 98 mm from the ignition). Overpressures generated from wires of 5 and 6 mm lengths are relatively low ($\Delta P_{max} = 0.290\,\text{bar}$ and $\Delta P_{max} = 0.343\,\text{bar}$, respectively) as compared to overpressures generated by greater wire lengths ($\Delta P_{max} = 0.488\,\text{bar}$ in average). The maximal overpressure is around 13% higher than maximum reached in the constant length tests with 6 mm (see previous section). It could be explained by the matter of rods used for the wire length study. Indeed, it is in pure copper whereas it is in copper-phosphorus (7%) alloy for the wire diameter and voltage study. The electric conductivity of rods being about 45% greater in case of pure copper (Davis 2001), then the energy delivered in the wire is more important in the wire length study. The rise time is about 0.0096 ms $\pm$ 0.0002 for all lengths, except for the 5 mm length where the rise time is 0.0103 ms. Therefore, the minimum optimum length to generate highest overpressures and shortest rise times is between 6 and 10 mm. From this minimal length, experimental values oscillate around mean values. These results are consistent with the main findings of Oktay (1965), that for an optimum discharge the initial voltage of the capacitor is directly proportional to the wire cross-sectional area and it does not depend on its length.

The ignition of a stoichiometric propane-oxygen mixture has been investigated. The wire diameter ($d = 0.23\,\text{mm}$) and the voltage ($U = 8\,\text{kV}$) were unchanged. Three wire lengths have been tested (6, 10, and 15 mm), leading to results in terms of

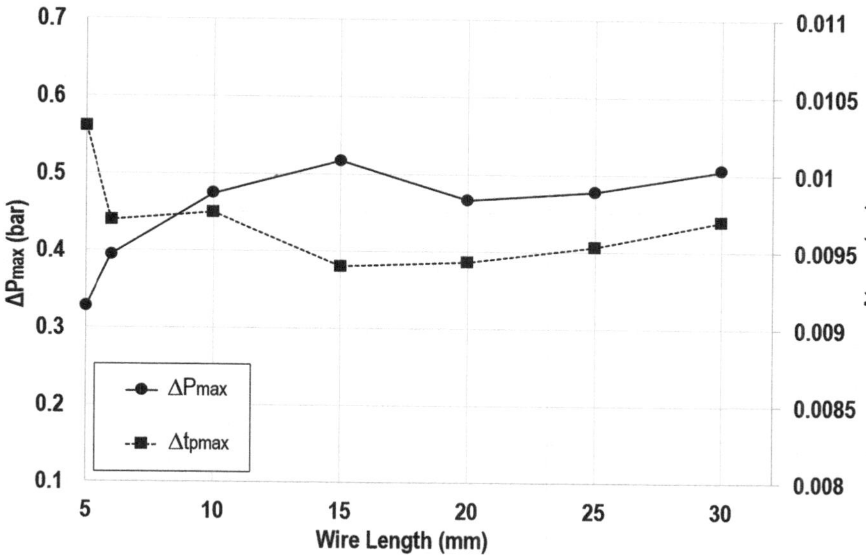

Fig. 5.5 Maximal overpressure and rise time according to the wire length from 98 mm to a copper wire explosion ($U = 8$ kV, $d = 0.23$ mm)

Table 5.2 Maximal overpressure measured at 98 mm from the ignition—Length effect

$C_3H_8 + 5\,O_2 - R = 47$ mm $- U = 8$ kV $- l = 6$ mm$- d = 0.23$ mm		
Length (mm)	$\Delta P_{max}/\Delta t_{pmax}$ wire (bar ms^{-1})	ΔP_{max} detonation (bar)
6	40.67	2.918
10	48.64	2.999
15	55.06	3.027

overpressure in Table 5.2. The measured pressure in case of the 6 mm long is lower than in the diameter study since the different height of the wire from the test table (18 mm vs. 8 mm) increases the wire—sensor distance of about 10 mm. As expected, only detonations have been observed for each testing condition since the ratio $\Delta P_{max}/\Delta t_{pmax}$ generated by the exploding wire at 98 mm from the ignition is over 23 bar ms^{-1} for all lengths of wire tested (see previous section), the minimum value being approximately 41 bar ms^{-1} for the 6 mm long. The maximal overpressure was obtained for a length of 15 mm ($\Delta P_{max} = 3.027$ bar), where the ratio $\Delta P_{max}/\Delta t_{pmax}$ is maximal (55.06 bar ms^{-1}).

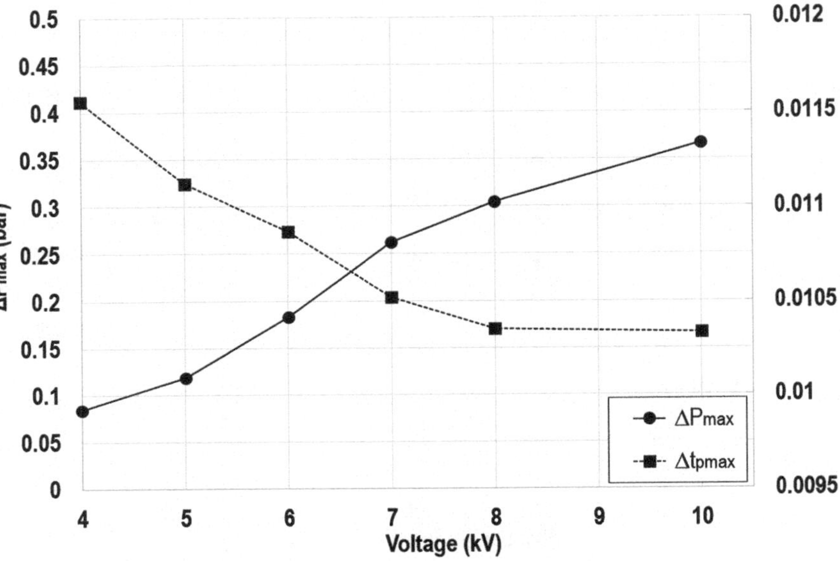

Fig. 5.6 Maximal overpressure and rise time according to the charge voltage U measured at 98 mm from the ignition ($d = 0.23$ mm, $l = 6$ mm)

5.3.3 *Influence of the Charge Voltage*

The effect of the charge voltage of the capacitor has been investigated. The charge voltages studied range from 4 kV up to 10 kV. The diameter has been fixed to $d = 0.23$ mm and the wire length has been kept to $l = 6$ mm. Measured maximal overpressures and rise times according to the charge voltage are reported in Fig. 5.6. It can be observed an almost linear increase in the maximal overpressure as a function of the voltage, to which it corresponds a linear decrease in the rise times. Indeed, the overpressure values increase from 0.084 bar to about 0.367 bar for a voltage of, 4 kV and 10 kV, respectively.

The effects of the charge voltage to ignite a stoichiometric propane-oxygen mixture were investigated, applying the previous methodology adopted for the wires exploding in free air keeping constant the wire diameter ($d = 0.23$ mm) and the wire length (6 mm). Experimental results are reported in Table 5.3. For a charge voltage of 5 kV (100 J) and 6 kV (144 J) both detonations and deflagrations can occur. The sample size of the experiments being not large enough to define a precise statistical trend, as a first approximation it can be inferred that the detonations at 5 kV and 6 kV occur around the 33% and 40% of the time, respectively.

As suggested in the diameter influence study, it seems that for a given combustible mixture, a critical value exists for the ratio $\Delta P_{max}/\Delta t_{pmax}$ generated by an exploding wire to ignite systematically a detonation. In this voltage influence study, the value of ratio $\Delta P_{max}/\Delta t_{pmax}$ is between 16 and 25 bar ms^{-1} (Fig. 5.7)

Table 5.3 Maximal overpressure measured at 98 mm from the ignition—Nominal energy effect

$C_3H_8 + 5\,O_2 - R = 47\,\text{mm} - d = 0.23\,\text{mm} - l = 6\,\text{mm}$

Voltage (kV)	Nominal stored energy (J)	$\Delta P_{max}/\Delta t_{pmax}$ wire (bar ms^{-1})	ΔP_{max} deflagration (bar)	ΔP_{max} detonation (bar)
4	64	7.28	0.218	–
5	100	10.68	0.348	2.787
6	144	16.84	0.559	2.991
7	196	24.90	–	3.024
8	256	29.42	–	3.259
10	400	35.59	–	3.422

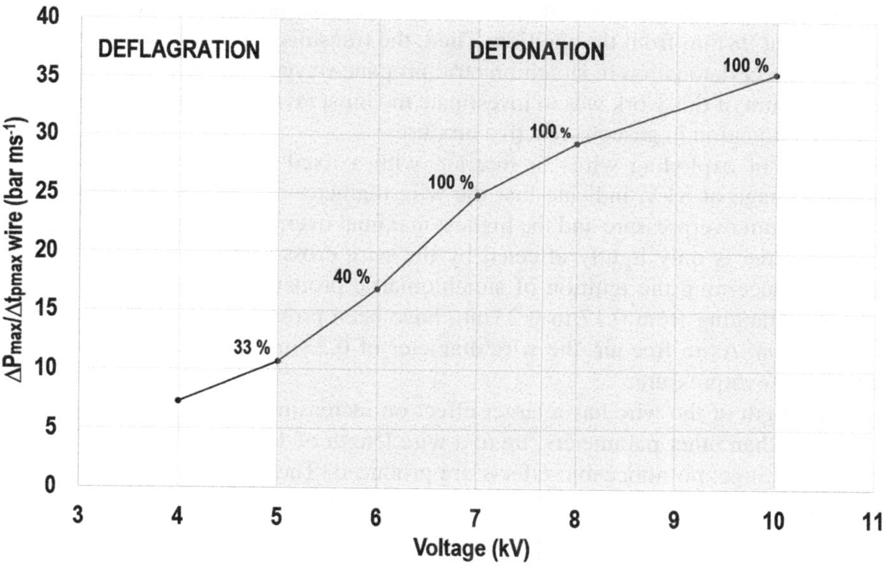

Fig. 5.7 Maximal overpressure—rise time ratio at 98 mm from the ignition according to the charge voltage U. Percent represents the probability to obtain a detonation ($d = 0.23$ mm, $l = 6$ mm, $R = 47$ mm)

for a stoichiometric propane-air mixture, at a distance of 98 mm from the ignition source. In the diameter influence study, this ratio value was evaluated to be in a smaller range, that is to say between 20.7 and 24.6 bar ms^{-1}, which is consistent with the previous result.

5.4 Conclusions

The first part of this study deals with the exploding wire phenomenon. A literature review presents processes that occur during the condensed wire fragments formation and the establishment of plasma around them. The formation of the first shock wave and sometimes the second shock wave is presented. Subsequently, investigations about the influence of several parameters on the exploding wire process, as the wire matter, diameter, length and the charge voltage of the capacitor were reported. Finally, the state of the art focuses on several studies performed with exploding wires in the goal to initiate a direct detonation in a reactive gaseous mixture.

An experimental parametric study with thin copper wires was performed at PRISME laboratory. The influence of the wire length l, wire diameter d, and charged voltage U was investigated on the pressure profile generated by an exploding wire, measured at 98 mm from the ignition. Then, the transmission of the shock wave to ignite a direct detonation in stoichiometric propane-oxygen mixture was observed. The main aim of this work was to investigate the most favorable conditions to ignite a direct detonation in gaseous reactive mixtures.

Results of exploding wires in free air, with a fixed length $l = 6$ mm and a charge voltage of 8 kV, indicate that the wire diameter of $d = 0.23$ mm presents the maximum overpressure and the highest maximal overpressure on rise time ratio. The rise time is only lightly affected by the wire cross-section (between 10 and 12 μs). Concerning the ignition of stoichiometric propane-oxygen mixture, wires diameters ranging from 0.17 to 0.23 mm have been proved to constantly generate a detonation. As in free air, the wire diameter of 0.23 mm generates the maximal detonation overpressure.

The length of the wire has a larger effect on increasing the overpressure profile in free air than other parameters, up to a wire length of 15 mm. However in the 15 to 30 mm range, no noticeable effects are produced. The same observation can be applied to the rise time. The $\Delta P_{max}/\Delta t_{pmax}$ ratio reaches a maximal value for the 15 mm long wire. The range of wire lengths going from 6 to 15 mm was always able to generate detonations in the propane-oxygen mixture with a charge voltage of 8 kV and a wire diameter of 0.23 mm.

The last part of this study concerns the influence of the charge voltage of the capacitor. In the 4 to 10 kV interval, an almost linear effect has been observed in increasing the maximum overpressure profile in free air and in decreasing its related rise time. The $\Delta P_{max}/\Delta t_{pmax}$ ratio follows the same trend. The minimum charge voltage which systematically ignites a detonation in a stoichiometric propane-oxygen mixture is 7 kV (196 J).

All tests performed in this work indicate that a minimum value of $\Delta P_{max}/\Delta t_{pmax}$ ratio is required to be certain to ignite a detonation in a stoichiometric propane-oxygen mixture. Indeed, it has been observed that the probability to generate a detonation increases with this ratio, from 33% when $\Delta P_{max}/\Delta t_{pmax} = 10.68$ to 100% when $\Delta P_{max}/\Delta t_{pmax} = 24.9$.

References

Anderson, J. A. (1922). The spectral energy distribution and opacity of wire explosion vapors. *Proceedings of the National Academy of Sciences of the USA, 8*, 231–232.

Bennett, F. D. (1958). Cylindrical shock waves from exploding wires. *Physics of Fluids, 1*(4), 347–352.

Brossard, J. (1970). Ondes de choc sphériques divergentes produites par l'explosion d'un fil métallique dans l'air. *Journal de Mécanique, 9*(3), 403–428.

Campanelli, M. (2016, September). Exploding wires. Master Internship Report, University of Orléans, Orléans, France.

Chace, W. G. (1959). Liquid behavior of exploding wires. *Physics of Fluids, 2*, 230.

Davis, J. R. (2001). *Copper and copper alloys*. Materials Park, OH: ASM International.

Hansen, S. P. (2011). *Exploding wires - Principles, Apparatus, and Experiments*. Owl's Head, ME: Bell Jar.

Knystautas, R., & Lee, J. H. (1976). On the effective energy for direct initiation of gaseous detonations. *Combustion and Flame, 27*, 221–228.

Lee, J. H. (1965, March). *The Propagation of shocks and blast waves in a detonating gas*. PhD Thesis, McGill University, Monreal 2, QC, Canada.

Lee, J. H., & Ramamurthi, K. (1976). On the concept of the critical size of a detonation kernel. *Combustion and Flame, 27*, 331–340.

Litchfield, E. L., Hay, M. H., & Forshey, D. R. (1963). Direct electrical initiation of freely expanding gaseous detonation waves. In *Ninth Symposium (International) on Combustion*. Cornell University Ithaca, New York, August 27 to September 1, 1962, Proceedings of the Combustion Institute (Vol. 9, Issue 1, pp. 282–286).

McGrath, J. R. (1966). Exploding wire. Research 1774–1963, U. S. NRL Memorandum Report: 1698.

Manson, N., & Ferrié, F. (1953). Contribution to the study of spherical detonation waves. In *Fourth Symposium (International) on Combustion*. Massachusetts Institute of Technology Cambridge, Massachusetts, September 1–5, 1952, Proceedings of the Combustion Institute (Vol. 4, Issue 1, pp. 486–494).

Nairne, E. (1774). Electrical experiments. *The Philosophical Transactions of the Royal Society (London), 64*, 79–89.

Nash, C. P., & Olsen, C. W. (1964). Initial phase of the exploding wire phenomenon. *Physics of Fluids, 7*(2), 209.

Oktay, E. (1965). Effect of wire cross section on the first pulse of an exploding wire. *Review of Scientific Instruments, 36*(9), 1327–1328.

Oshima, K. (1962). Blast waves produced by exploding wires. In *Exploding wires* (Vol. 2, pp. 159–174). New York: Springer.

Pikuz, S. A., Ivanenkov, G. V., Shelkovenko, T. A., & Hammer, D. (1999). On the phase state of the core matter in a high-power discharge through a wire. *JETP Letters, 69*, 377–382.

Romanova, V. M., Ivanenkov, G. V., Mingaleev, A. R., Ter-Oganesyan, A. E., Shelkovenko, T. A., & Pikuz, S. A. (2015). Electric explosion of fine wires: Three groups of materials. *Plasma Physics Reports, 41*(8), 617–636.

Sarkisov, G. S., Sasorov, P. V., Struve, K. W., & McDaniel, D. H. (2004). State of the metal core in nanosecond exploding wires and related phenomena. *Journal of Applied Physics, 96*(3), 1674–1686.

Schardin, H. (1954). Measurement of spherical shock saves. *Communications of Pure and Applied Mathematics, 7*, 223–243.

Taylor, M. J. (2002). Formation of plasma around wire fragments created by electrically exploded copper wire. *Journal of Physics D: Applied Physics, 35*(7), 700.

Turner, B. R. (1960). *A study of exploding wires*. PhD Thesis, California Institute of Technology, Pasadena, CA.

Volkov, L. P., Voronov, V. M., & Samylov, S. V. (1967). Some properties of shock waves produced in air by exploding wires. *Soviet Physics, JETP, 24*, 8–10.

Chapter 6
Blast Wave Experiments of Gaseous Charges

Isabelle Sochet and Yohann Maillot

Various approaches can be used to validate simulations (homemade or commercial codes) of the explosions and propagation of shock waves. The use of traditional explosives at large scale is largely used in the research centres of the field throughout the world. However, the safety requirements pyrotechnics, the price of the explosives and the sensors, the necessary availability of the sites and the duration of implementation of the tests (one to two tests per day to the maximum) do not make it possible to consider parametric experiments at a cost and a reasonable delay.

Several teams use shock tubes to generate shock in air for tests at medium or small scale. However, the traditional equipment does not make it possible to obtain the same characteristics of a shock wave generated by an explosion in air (peak of pressure followed by a brutal relaxation and a phase of depression). The current tubes with shock generate a maintained plate of overpressure (Ritzel and Parks 2010). Some of teams (Ritzel and Parks 2010; Chandra et al. 2012; Rae and Gunderson 2016; Maffeo et al. 2016) modify the geometry of high pressure chamber to generate pressure signals close to those of an explosion. However, in this case three defects remain: (a) the difficulty of simulating the explosion of various loads at various distances (because the only flexible physical parameter is the pressure of the high pressure chamber); (b) the scale necessarily very small of the configurations to be tested, which must be positioned in the tube or with the immediate vicinity of its end and (c) impossibility of generating non-plane shocks. Another teams use large blast tube open at two extremities and the explosive may be located anywhere within the length of the tube to generate blast wave with a typical Friedlander pressure decay (for example Robey 2001).

Small scale experiments present several advantages. Hence, the accuracy methods of laboratory can be applied, the tests conditions are perfectly controlled, the

I. Sochet (✉) • Y. Maillot
Laboratoire PRISME, INSA Centre Val de Loire, Bourges, France
e-mail: isabelle.sochet@insa-cvl.fr

I. Sochet (ed.), *Blast Effects*, Shock Wave and High Pressure Phenomena,
https://doi.org/10.1007/978-3-319-70831-7_6

reproducibility of tests is assured and it is easy to conduct parametric studies and to vary geometry of configurations and sizes of examined structures. Thus already existing works report small scale experiments conducted with high explosive charge. For example:

- Smith et al. (1998) used 72 g of PE4 with a model at 1:30 scale to investigate blast wave attenuation along tunnels in uniform cross-section with discrete roughness elements along the length
- Smith and Rose (2000) realised 1:50 scale experiments with SX2 explosive charge to investigate propagation of shock wave in urban geometry;
- Reichenbach and Neuwald (2001) conducted experimental studies in small-scale models using Nitropenta charges of 0.5 g to obtain a database for indoor detonations in a multi-chamber system
- Neuwald and Reichenbach (2002) characterised the blast effects at distance from the tunnel entrance with respect to a spherical and cylindrical charge of 1 g PETN and the location and orientation of this charge
- Reichenbach et al. (2002) realised a study on Mach reflection to validate the precision of laboratory experiments from spherical and cylindrical charge of 0.2–1.5 g of PETN
- Smith et al. (2004) examined the reduction in loading of blast that propagates from an array of domestic dwellings represented at 1:34 scale with a 18 g PE-4
- Miura et al. (2004) studied the reflection of the wave on the walls of a scaled model of nuclear fuel cycle facility. The explosive charge is 10 mg of silver azide exploded by laser beam
- Ripley et al. (2004a,b) used 50 g of C4 and 58 g sensitised nitromethane charges to investigate explosive blasts in urban scenarios using 1:24 scaled geometries
- Fouchier et al. (2017) presented an understanding of blast propagation in an urban environment at 1:200 scale with three types of charges: a pyrotechnic device (1.34 g of mass TNT equivalent); an RP-80 EBW detonator (0.136 g of mass TNT equivalent) and an RP-83 EBW detonator (1.31 g of mass TNT equivalent).

Moreover we must keep in mind the safety of experiments at small scale by using small explosive charges and the reduced cost.

In a latter chapter of this book, an original small scale experiment by means of exploding wire is presented by the team of G. Ben Dor and gives a good example of an application. These experiments consist of a knot isolated exploding wire between two electrodes. This setup allows generating spherical shock waves and the investigation of propagation and interaction of shocks with structures.

In this chapter, the methodology used by the laboratory PRISME is presented. It is based on the detonation of gas bubbles initiated electrically like means of reproducing on reduced scale the characteristics of a real explosion. This method does not induce damage with the structures located vicinity of the load (pressure CJ of about 20 bar), which makes it possible on the one hand to control the safety of the operators and on the other hand to produce inexpensive and light models (wood, polycarbonate, etc.) on varied scales. Equivalence gas/TNT is controlled for a large domain of reduced distance. Many configurations (unconfined and confined) are tested and showed the great versatility of this approach. Lastly, it is possible to carry out several shootings per day at a moderated cost, which authorises the achievement of parametric experimental campaigns to guide the developments of scale models.

The aim is to predict propagation of shock wave and loading applied on structures in unconfined or confined domain. The shock wave is generated by a strong explosion. Here, the explosion is a detonation of small gaseous charge. The analysis of shock wave provides analytical laws of overpressure, impulse, duration, arrival time versus depending on the simulated environment.

Properties of blast wave from small scale experiments are compared to results obtained at medium and large scale. They are realised with gaseous mixtures initially confined in hemispherical volume until 7 m^3 (Fraunhoffer institute), 134 m^3 (Dorofeev) and 510 m^3 (Brossard). Finally, the higher limit of experiments is proposed by Dewey with a volume of gas of 14479 m^3. Hence, the analysis of blast properties covers a large experimental domain from small to large scale.

6.1 Experimental Measurements : From Small to Large Scale

From 1960, the blast waves generated by hemispherical and spherical unconfined gaseous detonations have been studied. The behaviour of the shock wave is described by the characteristics of overpressure, arrival time, positive impulse and duration. In free field, the negative phase is often an overlooked aspect for potential hazards (Kogarko et al. 1965; Fishburn 1976; Baker et al. 1983; Zhdan 1983; Lannoy 1984; Borisov et al. 1985; Brossard et al. 1985; Gelfand et al. 1985; Dorofeev et al. 1993; Dorofeev 1995; Dewey 2005; Sochet and Schneider 2010). Here, some detailed works are selected to compare the characteristics of blast waves at different scales and to investigate the geometry of the gaseous charge (hemispherical and spherical).

6.1.1 Small Scale Experiments

6.1.1.1 Experimental Setup

The experimental setup is laboratory scale. The original concept is firstly described by Desbordes et al. (1978) and later by Brossard et al. (1988). It is based on detonation of gaseous charge. The experimental setup allows:

- Generating gaseous explosive charge with composition and volume governed by the user and perfectly known
- Supplying a controlled nominal energy

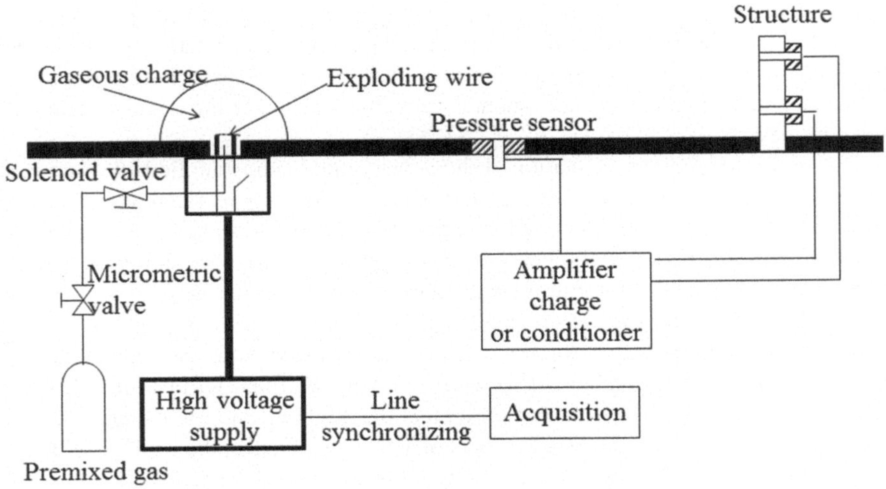

Fig. 6.1 Schematic of experimental setup for hemispherical charge

- Quantifying the characteristics of shock wave due to the gas detonation
- Ensuring the safety of the user and equipment.

All experiments are conducted at ambient pressure P_0 (1 bar) and temperature T_0 (293 K). The experimental setup is explained and illustrated in Figs. 6.1 and 6.2.

The test bench is a horizontal table on which gaseous charge and studied structure are placed. This table is a perfectly rigid and reflected plane.

In Figs. 6.1and 6.2, it is possible to distinguish the part of explosive and the part of metrology. The premixed gas can be injected under the table to create a hemispherical charge or above the table to obtain a spherical charge. The flow rate of the injected gas is controlled by a micrometric valve. Once the desired volume is achieved, a solenoid valve is closed and the bottle of premixed gas is disconnected. The igniter source is an exploding wire between two electrodes. These electrodes are connected to a high voltage supply. The exploding wire leads to the detonation of the gaseous charge. The igniter source (i.s.) is located at the centre of charge. However, some works concern the effect of igniter source location (Desrosier et al. 1991). The copper and zinc wire is 10 mm long and 0.16 mm diameter. More details on characteristics of other types of copper exploding wire are exposed in previous chapter section (Chap. 5).

The measurement chain is composed of pressure sensors (piezoelectric transducers, Kistler 603 B and PCB 113B26) and synchronous acquisition (1 MHz).

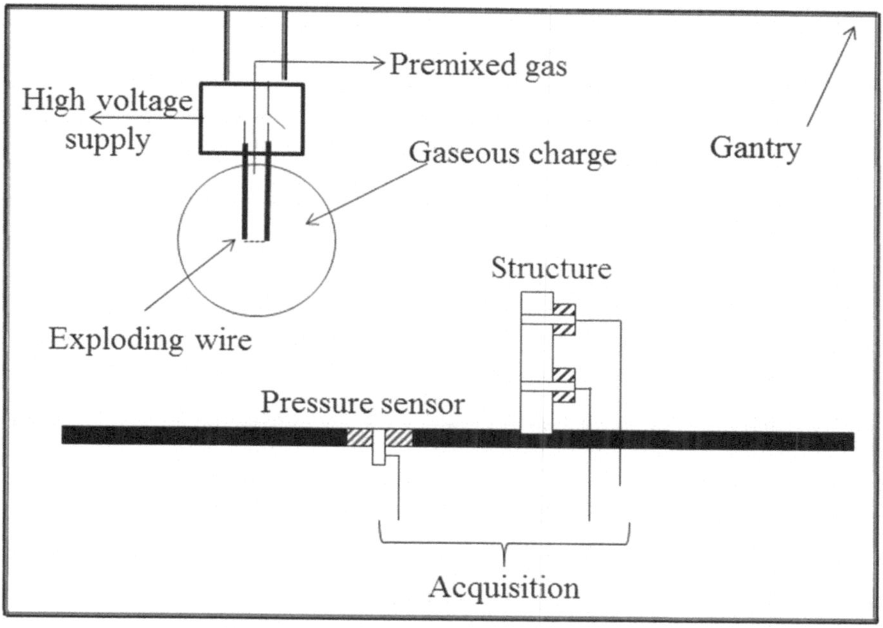

Fig. 6.2 Schematic of experimental setup for spherical charge

6.1.1.2 Gaseous Charge

The experiments are realised with gaseous mixture propane/oxygen in poor, stoichiometric or rich proportions. The premixed gas is first prepared in a 2 L bottle under 3 bar. This bottle is purged with inert gas before each filling. Proportion of gas is determined by partial pressure method. The gases are 99.99% pure.

The gaseous mixture can be confined in two types of envelopes depending on the considered study. Hence, it is possible to create hemispherical or spherical volume.

The hemispherical volume consists of half soap bubble. The confinement is created by injection of gas controlled by a micrometric valve. The diameter is limited by a metal ring placed on the surface of test bench. In free field, the hemispherical shape can be checked with a target. In case of spherical volume, a plastic balloon is used. The balloon is held in height by means of a loading gantry. The volume (i.e. diameter) is looked in two different ways. The first one is manual by using a ring. This ring is placed on the exterior envelope of the balloon and removed. The second one involves a picture. Hence, the spherical shape is controlled. The radiuses R_0 of the hemispherical and spherical gaseous charge vary between 0.03 and 0.10 m. The nature of envelope (soap bubble or plastic balloon) does not affect the explosion.

The energy released $\Delta H_R^{T_0}$ by the chemical reaction of the gaseous charge is calculated according to the enthalpy of formation $H_{Fi}^{T_0}$ at standard reference state (P_0, T_0) of each species (i) derived from Lide (2004).

For stoichiometric propane-oxygen ($C_3H_8 + 5O_2 \longrightarrow 3CO_2 + 4H_2O$) mixtures the energy released $\Delta H_R^{T_0}$ is $-2.044\,\mathrm{MJ\,mol}^{-1}_{C_3H_8}$. The energy E_v released by the detonation per unit of volume is determined by $E_v = \frac{|\Delta H_R^{T_0}|}{\sum_i n_i V_{mol}}$ where n_i represents the number of moles of the ith species in the mixture and V_{mol} the molar volume ($V_{mol} = 0.024\,\mathrm{m^3\,mol^{-1}}$ at P_0 and T_0).

The specific energy E is relative to spherical or hemispherical volume and calculated by $E = \frac{4}{3}\pi R_0^3 E_v$ or $E = \frac{2}{3}\pi R_0^3 E_v$, respectively.

The mass of gas mixture m is determined at (P_0, T_0) by the product of the mixture density ρ and the volume, such as $\rho = \frac{\sum_i n_i \mathrm{Mw}_i}{\sum_i n_i}\frac{P_0}{RT_0}$ with Mw_i the molecular weight mass of the ith species and the R universal gas constant ($8.314\,\mathrm{J\,mol^{-1}\,K^{-1}}$).

Table 6.1 brings data on the density ρ, the energy per unit volume E_v and per unit mass E_m for propane-oxygen mixtures at different equivalence ratio ϕ.

The specific energies and masses for stoichiometric propane-oxygen mixtures versus the initial radius R_0 are reported in Table 6.2, as an example.

6.1.1.3 Detonation of the Gaseous Charge

The detonation of the gaseous charge is ensured by an explosive wire releasing a nominal energy of 256 J. The detonation was checked by placing a pressure gauge inside the confinement of gaseous charge. The detonation process has been determined on the basis of Chapman-Jouguet pressure. For a stoichiometric propane-oxygen mixture, the Chapman-Jouguet pressure is in order of 36.7 bar as it is reported in the work of Schultz and Shepherd (2000). To ensure the detonation of the used mixture at small scale, a measurement of pressure is realised and reported here.

Table 6.1 Density and energy per volume for propane-oxygen mixtures versus equivalence ratio

C_3H_8/O_2			
ϕ	ρ (kg m^{-3})	E_v (MJ m^{-3})	E_m (MJ kg^{-1})
0.8	1.40	11.72	8.37
1.0	1.41	14.16	10.04
1.2	1.43	12.88	9.01

Table 6.2 Energy and mass for propane-oxygen mixtures versus radius of hemispherical charge

Hemispherical charge	C_3H_8/O_2	
R_0 (m)	E (MJ)	m (kg)
0.03	0.8×10^{-3}	0.08×10^{-3}
0.05	3.7×10^{-3}	0.37×10^{-3}
0.10	29.7×10^{-3}	2.96×10^{-3}

The explosive load consists of a hemisphere filled with a stoichiometric propane-oxygen gaseous mixture with a radius of 0.05 m. A pressure gauge is located inside the volume at 0.035 m from the centre of the ignition source. The maximum pressure measured is 30 bar and the time for the pressure rise is of 3 μs. Hence, the detonation wave evolves inside the reactive mixture until the boundary of air to reach Chapman-Jouguet state. This measure confirms that a detonation is really created inside the hemispherical gaseous charge.

The energy supply of 256 J is a nominal energy. Of course, the efficient energy is lower and represents 25–30% that corresponds to 50–60 J. The energy is negligible in relation to the energy released by the chemical reaction of ~3.7 kJ for a radius charge of 0.05 m.

The specific energies E are higher than the energy provided by the ignition source E_{is}. The ratio E/E_{is} is in order of 14–15. Hence, it means that the contribution of the initiation is negligible once the detonation is created.

The results at small scale experiments reported here are a compilation of all results obtained during the last decade at PRISME laboratory (S. Trélat, P.E. Sauvan, J. Baptiste, S. Eveillard, K. Gault, Y. Maillot). The results of these trials allow to collect important data with a good reproducibility. The mean of the standard deviation of arrival time is in order of 1%, that of overpressure is approximately of 3.5% and that of positive impulse and duration are in order of 5–8%. This last percent depends on the experimental configuration.

6.1.2 Medium Scale Experiments

The experiments conducted at medium scale correspond to the volumes of gas until 510 m^3 for hemispherical charge and 215 m^3 for spherical charge.

6.1.2.1 Experiments of Fraunhoffer Institute

The experiments realised at Fraunhoffer institute (Sochet and Schneider 2010) are conducted with stoichiometric ethylene-air mixtures (6.53% volumic proportions). The unconfined gas clouds are simulated by gas-filled hemispherical balloons of thin polyethylene envelopes. A solid explosive of 50 g Tetryl was used as a primer for the detonation of the gas mixtures and placed at the centre of the explosive charge. The volume of the hemispherical balloons ranges from 0.77 to 7.64 m^3 corresponding to masses of gas between 0.93 and 9.15 kg. The pressure-time history of the blast wave is recorded by 15 piezoresistive pressure sensors (Kistler 4041, 4071 and 4601) located at ground level between 5 and 50 m from the centre of the charges. Each characteristic (overpressure, arrival time, positive phase duration and positive impulse) of blast wave is analysed. At the time the tests were carried out the average atmospheric pressure was 0.993 bar. The characteristics of experimental conditions are reported in Tables 6.3 and 6.4.

Table 6.3 Density, energy per volume and mass for stoichiometric ethylene-air mixture—Experiments of Fraunhoffer

Mixture	ϕ	ρ (kg m^{-3})	E_v (MJ m^{-3})	E_m (MJ kg^{-1})
C_2H_4 + 14.28 Air	1	1.19	3.59	3.01

Table 6.4 Experiments of Fraunhoffer: experimental conditions

Charge radius R_0 (m)	Volume V (m^3)	Mass M (kg)	E (MJ)
0.718	0.777	0.93	2.80
1.539	7.642	9.15	27.53

Table 6.5 Density, energy per volume and mass for propane-air mixtures—Experiments of Dorofeev

Mixture	ϕ	ρ (kg m^{-3})	E_v (MJ m^{-3})	E_m (MJ kg^{-1})
C_3H_8 + 23.81 Air	1.00	1.22	3.42	2.80
C_3H_8 + 19.00 Air	1.25	1.23	3.23	2.63
C_3H_8 + 15.66 Air	1.52	1.24	3.03	2.45
C_3H_8 + 13.55 Air	1.76	1.24	2.83	2.27

6.1.2.2 Experiments of Dorofeev

Dorofeev presented air blast parameters (Dorofeev et al. 1993; Dorofeev 1995) obtained at medium scale with stoichiometric and rich propane-air mixtures. The gaseous mixtures are confined into plastic hemispheric envelope. The radius R_0 of charge is 4 m that is 134 m^3 and a mass of gas of 166 kg. This condition provides an energy released by chemical reaction in order of $E = 480$ MJ. The characteristics of experiments are given in Table 6.5. Two fans are used to ensure a homogeneous mixture. The detonation of gas is achieved by blasting HE charge in the centre of charge on the ground. The characteristics of blast are measured by 18 pressure gauges (Kistler 701 A) distributed on the ground along radial distances up 20 R_0. The authors reported only overpressure and positive impulse evolutions.

6.1.3 Experiments of Brossard

Brossard et al. (1985, 1988) reported blast waves characteristics obtained by unconfined gaseous detonations called AMEDE trials. This series of experiments was realised on the CESTA site in France between 1976 and 1982. Different gaseous mixtures are used. Here, only ethylene-air mixtures are retained. The mixture with 8 vol.% of ethylene (equivalence ratio of 1.24) is used for hemispherical volume. Spherical volumes are filled with 8 vol.% of ethylene and 16.2 vol.% (equivalence ratio equals to 2.76). The mixtures are initially confined in hemispherical or spherical volume limited by a mylar envelope. The uniformity of the gaseous medium is obtained by turbulence injection of gases and verified by two probes before each test. The ignition source is a solid explosive charge located at the centre of the volume. The pressure sensors are located on the ground level in

Table 6.6 Density, energy per volume and mass for ethylene-air mixtures—Experiments of Brossard

Mixture	ϕ	ρ (kg m^{-3})	E_v (MJ m^{-3})	E_m (MJ kg^{-1})
C_2H_4 + 11.50 Air	1.24	1.28	3.69	2.88
C_2H_4 + 5.71 Air	2.76	1.27	3.50	2.76

Table 6.7 Experiments of Brossard: experimental conditions

Geometry	HOB (m)	Charge radius R_0 (m)	Volume V (m^3)	Mass M (kg)	E (MJ)
HC	0	1.11	2.9	3.70	10.69
HC	0	1.92	14.9	19.10	54.93
HC	0	6.25	510.0	625.80	1880.06
SC	2.5	1.18	6.8	8.64	25.10
SC	3.2	1.18	6.8	8.64	25.10
SC	3.5	1.18	6.8	8.70	25.10
SC	4.5	1.18	6.8	8.64	25.10
SC	3.5	1.29	9.0	11.52	33.24
SC	3.5	1.85	27.0	34.29	94.57
SC	4.0	3.72	215.0	275.20	793.68

HC hemispherical charge, *SC* spherical charge

case of hemispherical charge and on the symmetry axis for spherical charges. Two types of piezoelectric transducers are used (Celesco LC 13 and Kistler). The team reported overpressure, positive impulse and time duration of the pressure signal. The characteristics of experiments are given below (Tables 6.6 and 6.7).

6.1.4 Large Scale Experiments

Dewey (2005) reported an experiment that is part of a series of tests conducted at Suffield, Canada and known as FE567 in Canada or as Operation Distant Plain in USA. The attention is focused on the detonation of a large hemispherical volume and the parameters of the resulting shock wave. The used gas was a mixture propane-oxygen in proportions of 1.43 for the equivalence ratio. The total mass of gas was 19,281 kg (~20 tons), that is 13,834 kg of oxygen and 5447 kg of propane or a mixture density of 1.33 kg m^{-3}. The diameter of the hemispherical gas-bag was 38.1 m, or a volume of 14,479.2 m^3. The atmospheric pressure was 0.94527 bar and the temperature at ground level was 321 K and at 1.83 m of altitude was 296.5 K. In this configuration, the volumic energy was 12 MJ m^{-3}, the mass energy was 9 MJ kg^{-1} and the total energy released by reaction is in order of 173,750 MJ.

Only arrival time at different distance was measured. The author has designed home made arrival detector (called ABTOAD for Air Blast Time of Arrival Detectors) and compared the measurements with high-speed photogrammetric methods. The shock radius is then expressed versus time of arrival by a fitted law which is differentiated to calculate de Mach number of the shock. The hydrostatic overpressure is obtained by applying Rankine-Hugoniot relationship.

These five scales of experiments used different gaseous mixtures (ethylene-air, propane-air and propane-oxygen). For hemispherical charge, the domain of energy released by chemical reaction is very large and covers the values from 0.8×10^{-3} MJ to 173,750 MJ, that is a ratio of 2×10^{8}.

6.2 Analysis of Blast Wave Characteristics

The four principal characteristics of blast wave are arrival time Ta, the positive overpressure ΔP, the positive phase duration T^{+} and the positive impulse I^{+}. These characteristics are classical expressed versus reduced radial distances. These distances are based on the principles of geometric similarity and conservation of momentum. Hopkinson (1915) and Sachs (1944) have introduced the scaled distances λ or Z by the following relations :

$$\lambda = \frac{R}{E^{1/3}} \tag{6.1}$$

or

$$Z = \frac{R}{M^{1/3}} \tag{6.2}$$

The term R represents the distance from the centre of explosion to the point of interest, E is the energy released by the chemical reaction and M the explosive mass, respecting the geometry of the charge. The scaled distances λ or Z are given in [m MJ$^{-1/3}$] and [m kg$^{-1/3}$], respectively.

We first examine the comparison between the effects generated by a spherical and hemispherical gaseous charge. Then the evolution of parameters is analysed versus reduced radial distances for a large range of scale size experiments described above. Arrival time, positive phase duration and positive phase are based on the cube root of energy or mass.

6.2.1 Comparison of Explosion Effects Generated by Hemispherical and Spherical Charges

It was noted in UFC (2008) that there is a difference on the evolution of incident overpressure versus reduced radial distance ($R.M^{-1/3}$) resulting from detonation of spherical and hemispherical TNT charge. The layout of the two curves is similar, but not parallel and the curve corresponding to the hemispherical explosive is above the spherical charge. So, we search to identify if a similar behaviour is obtained for gaseous charge and if it is dependent on the size scale experiments.

To compare the explosion effects created by hemispherical and spherical charges the measurements must be taken on the symmetry axis. In this topic, experiments were conducted. A radius $R_0 = 0.06$ m is used for the hemispherical and spherical charge and this last one is located at altitude HOB $= 0.3$ m. The two explosive charges are filled with a stoichiometric propane-oxygen mixture. Pressure gauges are distributed on the ground at different distances between $R = 0.1$ and 2.3 m from the centre of the charge for the hemispherical charge scenario. In case of the configuration of spherical volume, the sensors are mounted on lollipops at the same height than the centre of ignition (HOB $= 0.3$ m) over distances between $R = 0.157$ and 1.570 m. For each configuration, the pressure sensors were PCB 113B26.

Figures 6.3 and 6.4 show a comparison between the two configurations of wave diagram and incident overpressure versus distance. It results that the distances of the propagation of shock wave versus arrival time are superimposed for the two cases. As can be seen in Fig. 6.4, the two curves are slightly different for the distances lower than 3 m and the overpressure resulting of spherical explosive detonation is higher than a hemispherical charge. Hence, through these experiences the mechanical effects induced by a spherical gaseous charge could be comparable but slightly more severe than generated by hemispherical charge.

However, pressure profiles differ according to the geometry of the charge if the sensors are not placed on the axis of symmetry of spherical charge. An incident shock wave propagates when a hemispherical charge detonates on the ground in free field. It is in contrast to the complex of shock waves creating by the explosion

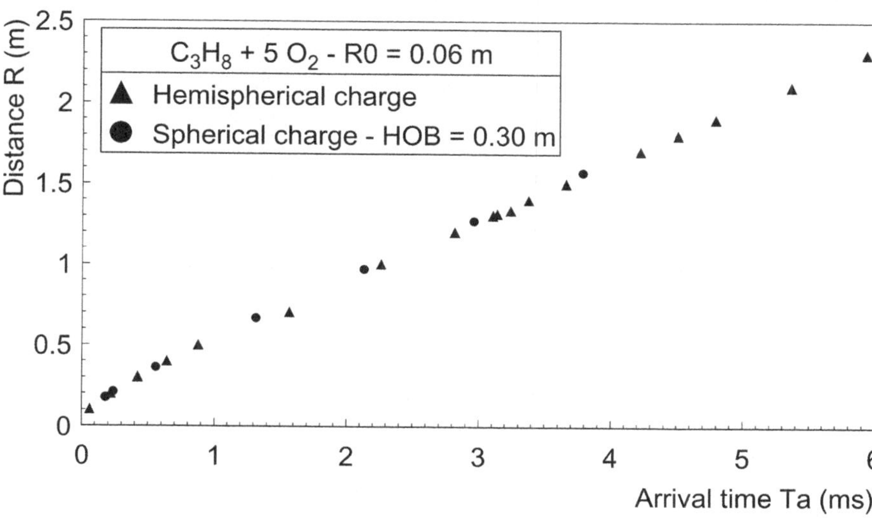

Fig. 6.3 Measured arrival time of the primary shock and effect of geometry of the gaseous charge

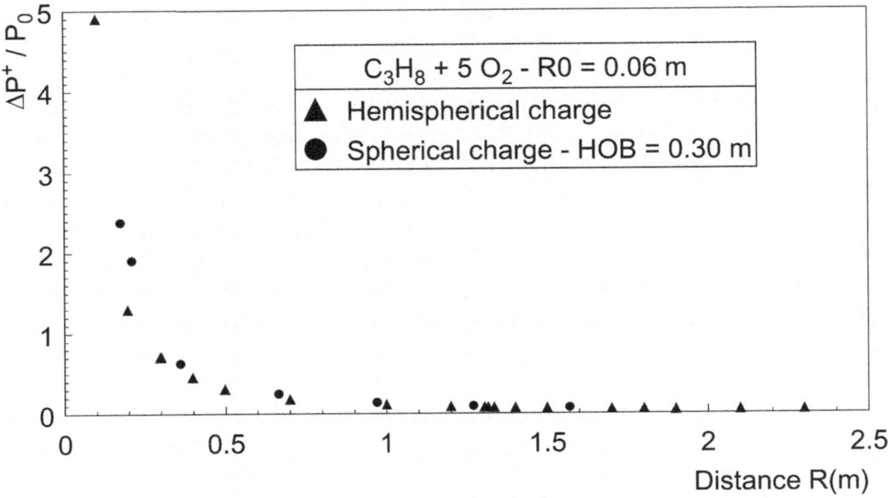

Fig. 6.4 Measured overpressure versus radius and effect of geometry of the gaseous charge

of a spherical explosive above the ground. In this last case, the incident shock wave is reflected on the ground and a transition from regular to irregular reflection can appear with a development of Mach Stem. The shock wave propagation is visualised to explain the evolution of pressure profile recorded during the detonation of the spherical gaseous charge ($R_0 = 0.06$ m, HOB $= 0.3$ m). A retro-reflective-shadowscopy setup was used as presented by Hargather and Settles (2009) and the camera was Phantom V7.3. The visualisation domain contains two pressure sensors located at ground level and, respectively, at 0.665 and 0.970 m from the centre of the initial gaseous charge. The pressure profiles are given in figure 6.5.

It can be observed that the incident shock wave approached gauge A at $t = 1.331$ ms (Fig. 6.6) and arrived on the gauge at $t = 1.366$ ms (Fig. 6.5) to reach gauge B at $t = 2.187$ ms or some later than it can be seen from photo. The discrepancy could be due to the orientation of camera and the parallax error. The reflected shock reaches the gauge A at 1.962 ms and gauge B at 2.596 ms, its corroborates the photos (Figs. 6.5, 6.7 and 6.8). Hence, the analysis of video and pressure signals sustains the reflection of shock wave on the ground and propagation. This result confirmed the capability of small scale experiments to study blast wave propagation.

The examination of the incident overpressure on a range of reduced radial distance obtained at medium scale is inconsistent with this observation. Indeed, the experiments conducted by Brossard et al. (1985) showed that fitted curves of overpressure resulting from detonation of hemispherical and spherical charges are not superimposed (Fig. 6.9). The evolutions are not parallel and so a coefficient of $2^{1/3}$ cannot explain the difference between two geometries of charges. Besides, the

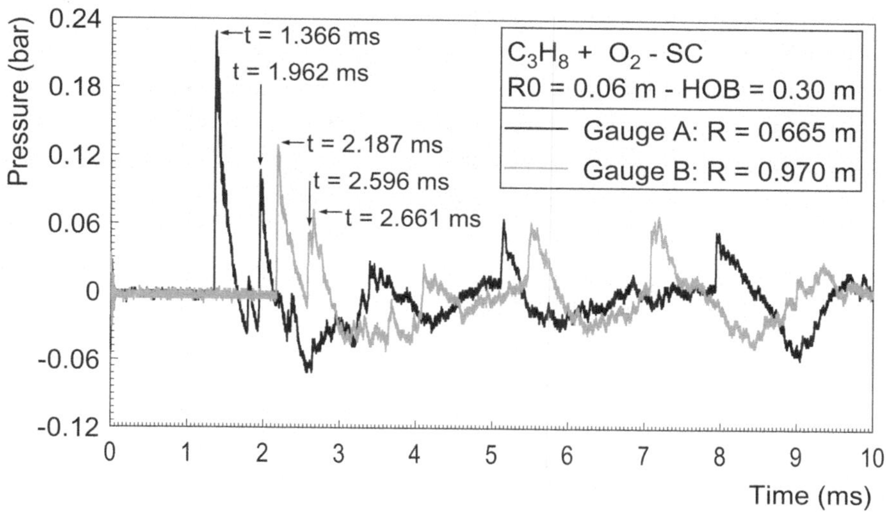

Fig. 6.5 Recorded pressure profiles at different distances for the detonation of spherical gaseous charge above the ground

Fig. 6.6 Shock wave propagation at $t = 1.331$ ms—Detonation of propane-oxygen charge of radius $R_0 = 0.06$ m at HOB = 0.3 m

Fig. 6.7 Shock wave propagation at $t = 1.931$ ms—Detonation of propane-oxygen charge of radius $R_0 = 0.06$ m at HOB = 0.3 m

Fig. 6.8 Shock wave propagation at $t = 2.568$ ms—Detonation of propane-oxygen charge of radius $R_0 = 0.06$ m at HOB = 0.3 m

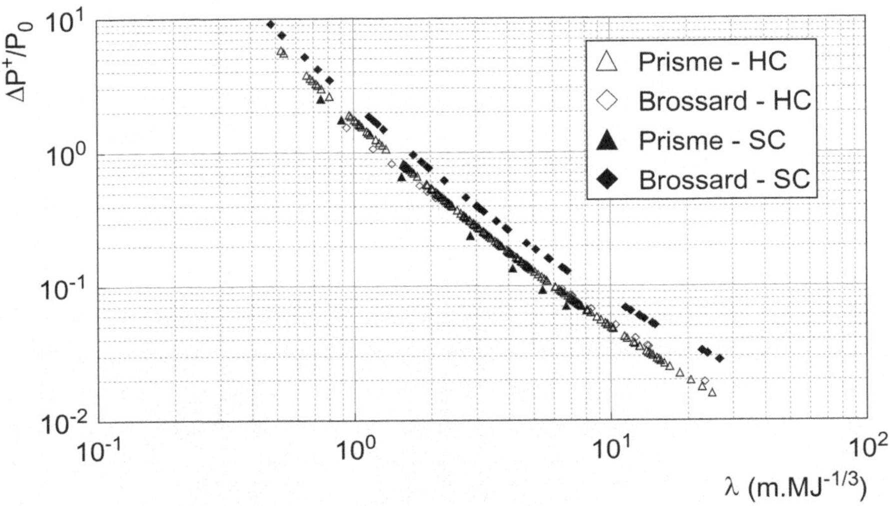

Fig. 6.9 Overpressure versus scaled distance in energy and effect of size and geometry of gaseous charges

overpressure associated with the spherical volume is higher than those created by hemispherical volume detonation. This point is opposite of behaviour observed with TNT charges (UFC 2008).

The atmospheric conditions and the nature of the soils during trials are not indicated by the authors. These factors impact the dissipation of energy. Instead, small scale experiments are conducted with a perfectly reflective ground and approximatively constant environmental conditions.

This study suggests that should be recommended to express all parameters of blast wave versus distance always respecting the geometry of the explosive charge. It is apparent that the factor $2^{1/3}$ could not be systematically applied. The relationships given in literature are often expressed for spherical volume. Then they should be used with caution. This analysis confirmed the capability of small scale experiments to characterise arrival time and overpressure of blast wave propagation generated by hemispherical or spherical gaseous charge.

6.2.2 Evolution of Characteristics of Blast Waves

Here, we consider the results of experiments described above. The diagrams reported in this section take into account the geometry used during the experiments of the authors. It means that the results published by Brossard et al. (1985), Dorofeev et al. (1993), Sochet and Schneider (2010), Trélat et al. (2007), Sochet et al. (2014, 2017), Julien et al. (2016), Maillot et al. (2017) and Gault (2017) have been converted with the appropriate volume of charges. The comparison of the characteristics parameters of blast wave is given versus the scaled distances λ. In addition, all polynomials laws are given in Appendices A and B of this book for a safety approach (properties of blast wave versus scaled distance) and for a forensic determination (scaled distance versus properties of blast wave) for the two scaled distances λ and Z.

There is an important effect of large scale size of experiments (Dewey 2005) on the arrival time of shock wave on pressure gauge (Fig. 6.10) in near field, i.e. $\lambda \prec 1\ \mathrm{m\,MJ^{-1/3}}$. In far field, i.e. $\lambda \succ 10\ \mathrm{m\,MJ^{-1/3}}$, there is convergence of arrival type between the three sizes of experiments (Prisme, Fraunhoffer and Dewey). A very small deviation is noted between experiments from Prisme and Fraunhoffer in the range $\lambda \prec 1.7 - 7\,\mathrm{m\,MJ^{-1/3}}$. However, arrival time from Prisme and Fraunhoffer is identical for the spherical charge (Fig. 6.11). Hence, the arrival time is independent of gaseous mixture, size and geometry of charges in far field of explosive charge.

The evolution of the overpressure versus the scaled distance in energy (Fig. 6.12) indicates very similar decay, except for tests of Fraunhoffer institute. The curves (Prisme, Dorofeev 1995; Brossard et al. 1985; Dewey 2005) overlap on λ range from 2 to $10\,\mathrm{m\,MJ^{-1/3}}$. The experiments with a hemispherical volume of 14,479.2 m^3 with propane-oxygen (Dewey 2005) indicate more important overpressure at near field and far field. It could be explained by the calculated pressure (via Rankine-

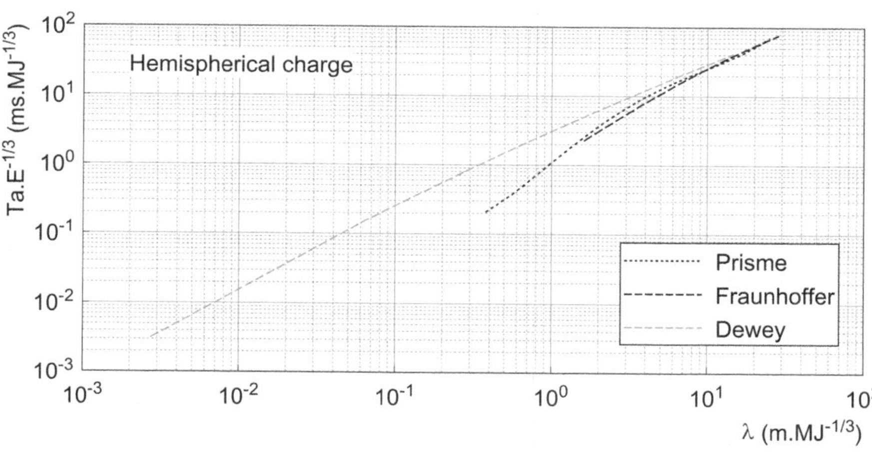

Fig. 6.10 Reduced arrival time versus scaled distance in energy for hemispherical gaseous charge

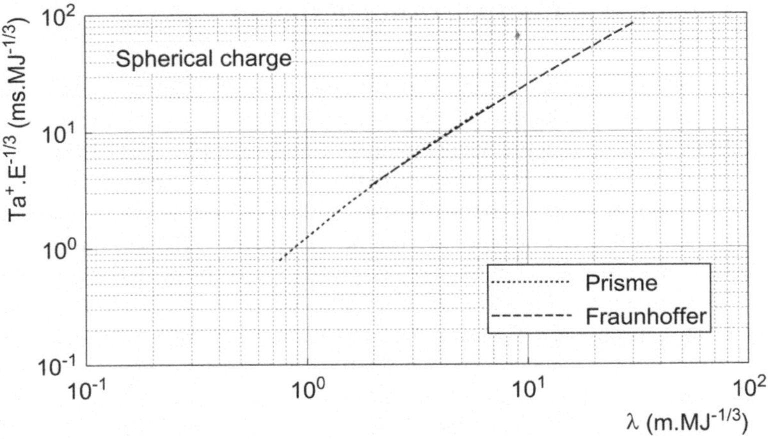

Fig. 6.11 Reduced arrival time versus scaled distance in energy for spherical gaseous charge

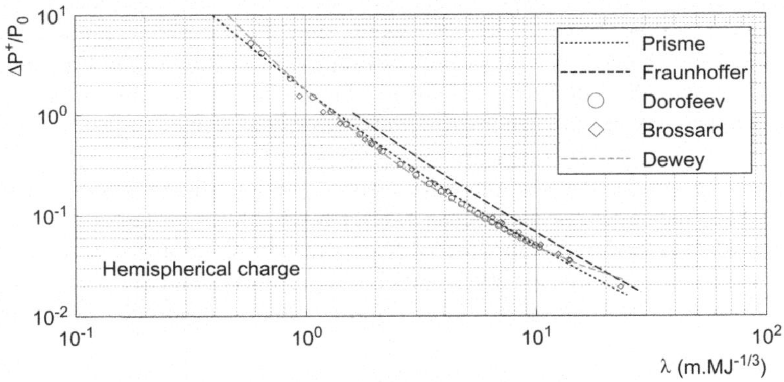

Fig. 6.12 Overpressure versus scaled distance in energy for hemispherical gaseous charge

Hugoniot from measurements of arrival time) unlike pressure measurements with other authors as it has been highlighted in Chap. 4. In case of spherical volume (Fig. 6.13) the discrepancy between small scale and medium scale is amplified. First, with respect to the experiments of Fraunhoffer the overpressure is clearly higher than the trials of Brossard. It is related to the position of pressure gauges at ground level and not at the same HOB than the centre of charge. Consequently, the pressure can be a reflected pressure and so more important than an incident one. As we explained in previous section, the incident pressure is smaller in case of spherical charge than a hemispherical one at small scale. The curves are parallel and the ratio of scaled distance ($\lambda_{SC}/\lambda_{HC}$) for a given overpressure is in average 0.9. As considering previously for the experiments of Brossard, the ratio $\lambda_{SC}/\lambda_{HC}$ is higher than one and higher than $2^{1/3}$.

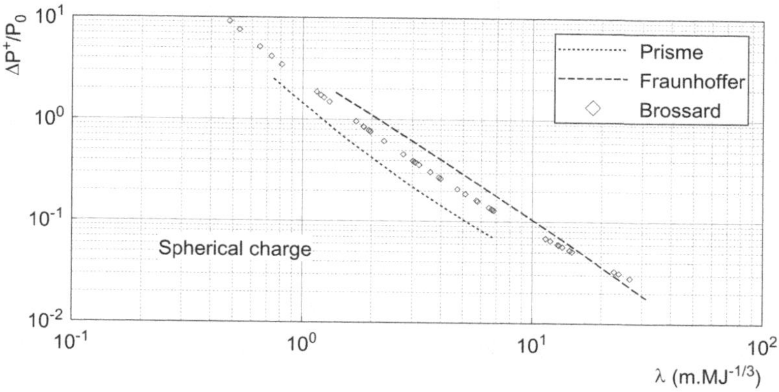

Fig. 6.13 Overpressure versus scaled distance in energy for spherical gaseous charge

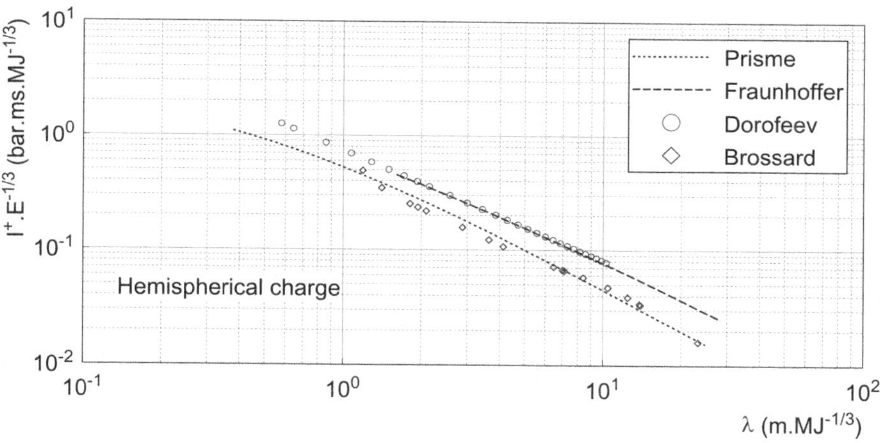

Fig. 6.14 Reduced positive impulse versus scaled distance in energy for hemispherical gaseous charge

Examination of Fig. 6.14 shows two groups of results, one defined by Fraunhoffer and Dorofeev results and the second one by Prisme and Brossard. The positive impulse obtained during the experiments of Dorofeev and those of Fraunhoffer overlaps on a scaled distance interval and in a perfect continuity for the opposite limits. As can be seen from Fig. 6.15 the positive phase duration reduced to energy is close to each other.

The results of impulse and positive duration for spherical volume are not expressed on fitted laws due to the dispersion of experimental measurements (Figs. 6.16 and 6.17). The more important disparity in the results for the positive impulse and duration is explained by the difficulty to determine these two parameters with a high accuracy. This determination depends on the quality of the pressure

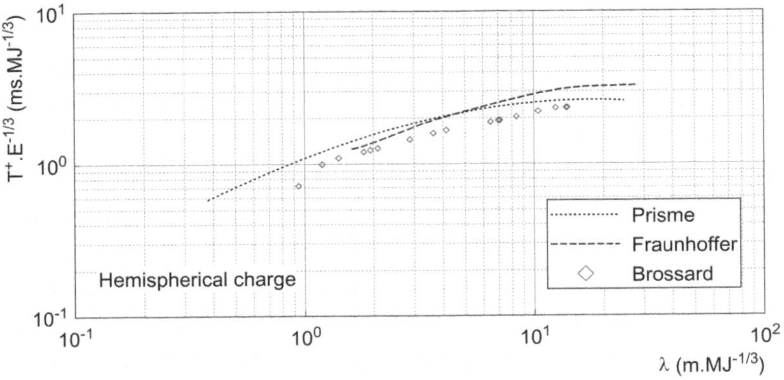

Fig. 6.15 Reduced positive phase duration versus scaled distance in energy for hemispherical gaseous charge

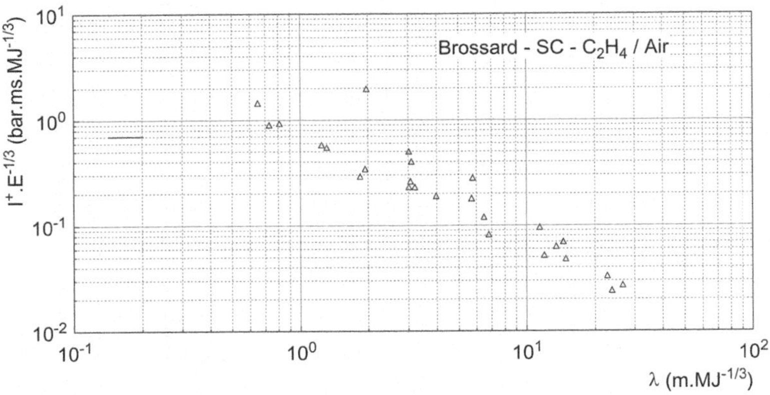

Fig. 6.16 Reduced positive impulse versus scaled distance in energy for spherical gaseous charge—experiments from Brossard

signal and the methodology applied to analyse it. In addition, for spherical charge, the reflected shock wave from the ground can be recorded on the pressure signal. If the peak reflected pressure is produced during the positive phase of incident wave, then the integral of the pressure in time is increased. In this case, this signal cannot be compared to a pressure signal generated by an incident shock wave.

The results obtained at small scale are efficient and reveal this interest of these experiments. The analysis provides that small scale studies:

- of the peak hydrostatic overpressure are in good correlation with larger experiments expressed versus λ or Z
- the arrival time is consistent with Fraunhoffer institute results on the entire field and with Dewey experiments at far field

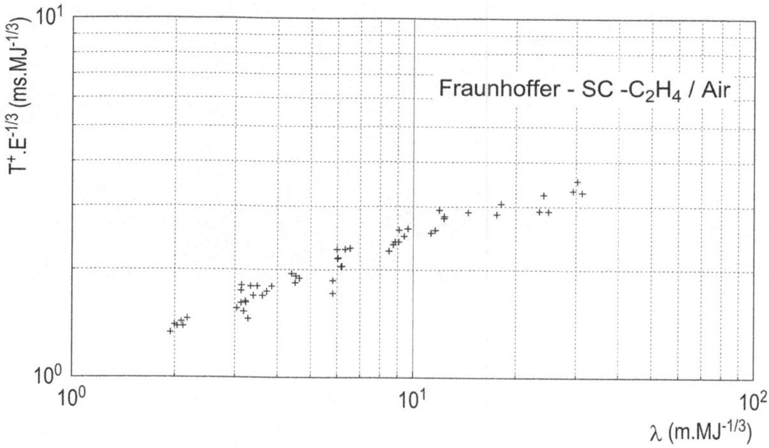

Fig. 6.17 Reduced positive phase duration versus scaled distance in energy for spherical gaseous charge—experiments from Fraunhoffer

- positive impulse versus λ is correlated with Brossard experiments on the same range scaled distance
- and evolution of positive duration could be compared with Fraunhoffer institute and Brossard experiments.

Hence, in this context small scale experiments have been conducted to investigate shock wave propagation and reflected shock in complex environment which can be adapted to industrial or urban scenarios (Trélat et al. 2007; Sochet et al. 2014, 2017; Julien et al. 2016; Maillot et al. 2017; Gault 2017).

6.2.3 Secondary Shock

The presence of a secondary shock appears on the pressure profile of blast wave, in some cases. This secondary shock phenomenon has been defined by Baker et al. (1983), Needham (2010), and Dewey (2001) and has been studied by Gitterman (2014). The authors explained that the detonation of gaseous charge generates a primary shock wave and an unstable contact surface between detonation products and air. The deceleration of this contact surface produces a rarefaction wave which moves outward and an inward-moving second shock. The second shock implodes and is reflected outward. Then other interactions with contact surface can take place.

Gitterman analysed the second shock through large-scale explosions conducted by the Geophysical Institute of Israel at the Sayarim Military Range in the Negev desert. Two mass of high-explosives were used: 82 and 100 tons. The second shock was found easily to measure and important to consider in properties of shock waves.

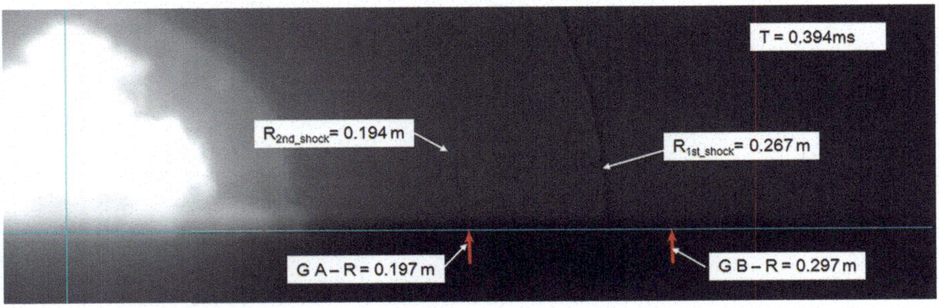

Fig. 6.18 Visualisation of primary and secondary shock waves at $t = 0.394$ ms—detonation of propane-oxygen hemispherical charge of radius $R_0 = 0.05$ m

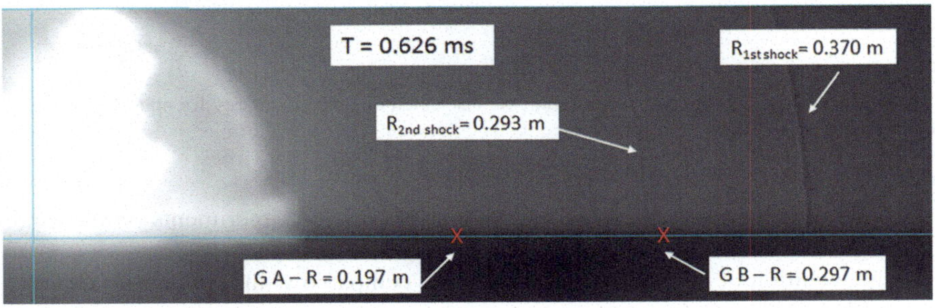

Fig. 6.19 Visualisation of primary and secondary shock waves at $t = 0.626$ ms—detonation of propane-oxygen hemispherical charge of radius $R_0 = 0.05$ m

In this context, an investigation of secondary shock has been realised at small scale. For that, a hemispherical charge of $C_3H_8 + 5O_2$ mixture was considered with a radius of 0.05 m. Two pressure gauges A and B were located at 0.197 and 0.297 m from the centre of charge, respectively. Simultaneously, a video was taken with a high speed camera. The visualisation setup is the same as the one used to compare the explosion effects generated by hemispherical and spherical charges. The propagation of primary and secondary shock can be observed at time $T = 0.699$ ms and $T = 0.626$ ms (Figs. 6.18 and 6.19). The pictures can be correlated with the pressure profile (Fig. 6.20). Hence, the incident shock arrived at $R = 0.197$ m at time 0.254 ms and the second shock swept the sensor at time 0.425 ms. The first photo corresponds to a time after the second shock. The shock wave propagated and reached the gauge B at 0.483 ms and followed by the secondary shock at $T = 0.691$ ms. The visualisation corroborates the pressure signal and the presence of secondary shock at small scale.

Note 1 In Appendices A and B of this book the reader will find all polynoms laws to calculate the physical properties of blast waves versus scaled distance expressed in terms of energy and mass. The reverse laws are equally reported.

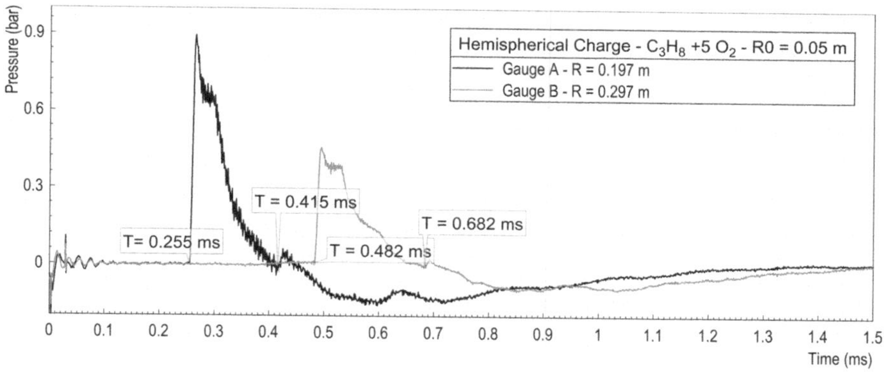

Fig. 6.20 Recorded pressure profiles at 0.197 and 0.297 m for the detonation of hemispherical gaseous charge

References

Baker, W. E., Cox, P. A., Westine, P. S., Kulesz, J. J., & Strehlow, R. A. (1983). *Explosion hazards and evaluation. Fundamental studies in engineering*. New York: Elsevier.

Borisov, A. A., Gelfand, B. E., & Tsyganov, S. A. (1985). On modelling of pressure waves formed by detonation and combustion of gas mixtures. *Journal of Physics of Combustion and Explosion (Russian Academy of Sciences), 2*, 90.

Brossard, J., Bailly, P., Desrosier, C., & Renard, J. (1988). Overpressures imposed by a blast waves. In *Progress in astronautics and aeronautics* (Vol. 114, pp. 389–400). Washington, DC: AIAA.

Brossard, J., Leyer, J. C., Desbordes, D., Saint-Cloud, J. P., Hendricks, S., Garnier, J. L., et al. (1985). Air blast from unconfined gaseous detonation. In *Progress in astronautics and aeronautics* (Vol. 94, pp. 556–566). Washington, DC: AIAA.

Chandra, N., Ganpule, S., Kleinschmit, N. N., Feng, R., Holmberg, A. D., Sundaramurthy, A., et al. (2012). Evolution of blast wave profiles in simulated air blasts: Experiment and computational modeling. *Shock Waves 22*(5), 403–415.

Desbordes, D., Manson, N., & Brossard, J. (1978). Explosion dans l'air de charges sphériques non confinées de mélanges réactifs gazeux. *Acta Astronautica, 5*(11–12), 1009–1026.

Desrosier, C., Reboux, A., & Brossard, J. (1991). Effect of asymmetric ignition on the vapor cloud spatial blast. In *Progress in astronautics and aeronautics* (Vol. 134, pp. 21–37). Washington, DC: AIAA.

Dewey, J. M. (2001). Spherical shock waves: Chapter 13.1 expanding spherical shocks (blast waves). In G. Ben-Dor, O. Igra, & T. Elperin (Eds.), *Handbook of shock waves*. Amsterdam: Elsevier.

Dewey, J. M. (2005). The TNT equivalence of an optimum propane-oxygen mixture. *Journal of Physics D: Applied Physics, 38*, 4245–4251.

Dorofeev, S. B., Sidorov, V. P., Dvoinishnikov, A. E., Alekseev, V. I., & Kyznetsov, M. S. (1993). Experimental study of air blast parameters from fuel rich mixtures detonation. In *14th International Colloquium on Dynamics of Explosions and Reactive Systems (ICDERS), 2/E1.4.1*.

Dorofeev, S. B. (1995). Blast effects of confined and unconfined explosions. In *20th International Symposium on Shock Waves, Pasadena, CA, USA* (Vol. 1, pp. 77–86).

Fishburn, B. D. (1976). Some aspects of blast from fuel-air explosives. *Acta Astronautica, 3*, 1049.

Fouchier, C., Laboureur, D., Youinou, L., Lapebie, E., & Buchlin, J. M. (2017). Experimental investigation of blast wave propagation in an urban environment. *Journal of Loss Prevention in the Process Industries*. https://doi.org/10.1016/j.jlp.2017.06.021

Gault, K. (2017). Private report for DGATN.

Gelfand, B. E., Gubin, C. A., Mikhalkin, V. N., & Shargatov, V. A. (1985). Computation of shock wave parameters by detonation of combustible gaseous mixtures of a variable composition. *Journal of Physics of Combustion and Explosion (Russian Academy of Sciences), 3*, 92.

Gitterman, Y. (2014). Secondary shock features for large surface explosions: Results from the Sayarim Military Range, Israel and other experiments. *Shock Waves, 24*, 267–282.

Hargather, M. J., & Settles, G. S. (2009). Retroreflective shadowgraph technique for large-scale flow visualization. *Applied Optics 48*(22), 4449–4457.

Hopkinson, B. (1915). British Ordnance Board Minutes, 13565.

Julien, B., Sochet, I., & Vaillant, T. (2016). Impact of the volume of rooms on shock wave propagation within a multi-chamber system. *Shock Waves, 26*(2), 87–108.

Kogarko, S. M., Adushkin V. V., & Liamin, A. G. (1965). Investigation of spherical detonations in gas mixtures. *Combustion, Explosion, and Shock Waves USSR, 1*, 2–22.

Lannoy, A. (1984). Analyse des explosions air-hydrocarbure en milieu libre : Etudes déterministes et probabiliste du scénario d'accident. Prévision des effets de surpression, Bulletin Direct. Etudes et Recherche EDF. A4 (Published in French).

Lide, D. R. (Ed.), (2004). *Handbook of chemistry and physics* (84th ed.). Boca Raton: CRC Press.

Maffeo, M., Carboni, M., Cyganik, J., Decristofano, B., Carneal, C., Zinn, D., et al. (2016). Test method development for the evaluation of head borne equipment with a blast simulator. In *24th International Symposium on Military Aspect of Blast and Shock (MABS 24), Halifax, Canada.*

Maillot, Y., Sochet, I., Vinçont, J. Y., & Grillon, Y. (2017). Etude expérimentale de la réflexion de Mach, 23ème Congrès Français de mécanique, Lille, France (Published in French).

Miura, A., Matsuo, A., Mizukaki, T., Shiraishi, T., Utsunomiya, G., Takayama, K., et al. (2004). Reflection and diffraction phenomena of blast wave propagation in nuclear fuel cycle facility. *Japan Society Mechanical Engineering, 47*(2), 287–292.

Needham, C. E. (2010). *Blast waves*. Berlin: Springer.

Neuwald, P., & Reichenbach, H. (2002). Detonations in front of a tunnel en-trance: A parametric small-scale study. In *17th International Symposium on Military Aspect of Blast and Shock (MABS 17), Las Vegas, NV, USA*

Rae, P. J., & Gunderson, J. (2016). Characterization of a large shock tube. In *24th International Symposium on Military Aspect of Blast and Shock (MABS24), Halifax, Canada.*

Reichenbach, H., & Neuwald, P. (2001). Indoor detonations - visualization and pressure measurement in small-scale models. In *24th International Congress on High-Speed Photography and Photonics, Proceedings of SPIE* (Vol. 4183, pp. 92–104). ISBN 9780819438461.

Reichenbach, H., Neuwald, P., & Kuhl, A. L. (2002). Role of precision laboratory experiments in the understanding of large-scale blast phenomena. In *Julius J. Meszaras Lecture, 17th International Symposium on Military Aspect of Blast and Shock (MABS17), Las Vegas, NV, USA.*

Ripley, R. C., Dunbar, T. E., Donhaue, L., & Von Rosen, B. (2004a). Personnel vulnerability predictions using small-scale air blast modeling. In *18th International Symposium on Military Aspect of Blast and Shock (MABS18), Bad Reichenhall, Germany*

Ripley, R. C., Von Rosen, B., Ritzel, D. V., & Whitehouse, D. R. (2004b). Small-scale modeling of explosive blasts in urban scenarios. In *21st International Symposium on Ballistics, Adelaide, Australia.*

Ritzel, D. V., & Parks, S. A. (2010). Blast simulation using shock-tube technology. In *21st International Symposium on Military Aspect of Blast and Shock (MABS21), Jerusalem, Israel.*

Robey, R. (2001). Blast tubes, Chapter 4.4. In G. Ben-Dor, O. Igra, & T. Elperin (Eds.), *Handbook of shock waves* (Vol. 1). New York: Academic Press.

Sachs, R. G. (1944). The dependence of blast and ambient pressure and temperature. BRL Report 466.

Schultz, E., & Shepherd, J. (2000). Validation of detailed reaction mechanisms for detonation simulation. Explosion Dynamics Laboratory Report FM99-5. http://authors.library.caltech.edu/25820/1/FM99-5.pdf

Smith, P. D., Vismeg, P., Teo, L. C., & Tingey, L. (1998). Blast wave transmission along rough-walled tunnels. *International Journal of Impact Engineering, 21*(6), 419–432.

Smith, P. D., & Rose, T. A. (2000). Influence of urban geometry on blast wave resultants. In *16th International Symposium on Military Aspect of Blast and Shock (MABS16), Oxford, England.*

Smith, P. D., Rose, T. A., & Ng, S. H. (2004). The influence of areal density on the shielding and channeling of blast by buildings. In *18th International Symposium on Military Aspect of Blast and Shock (MABS18), Bad Reichenhall, Germany.*

Sochet, I., & Schneider, H. (2010). Blast wave characteristics and equivalency. In S. M. Frolov, F. Zhang, & P. Wiolanski (Eds.), *Explosion, dynamics hazards* (pp. 169–184). Moscow: Torus Press.

Sochet, I., Sauvan, P. E., Boulanger, R., & Nozeres, F. (2014). Effect of a gas charge explosion at the closed end of a gas storage system. *Journal of Loss Prevention in the Process Industries, 27*, 42–48.

Sochet, I., Eveillard, S., Vinçont, J. Y., Piserchia, P. F., & Rocourt, X. (2017). Influence of the geometry of protective barriers on the propagation of shock waves. *Shock Waves, 27*(2), 209–219.

Trélat, S., Sochet, I., Autrusson, B., Cheval, K., & Loiseau, O. (2007). Strong explosion near a parallelepipedic structure. *Shock Waves, 16*(4–5), 349–357.

U.S. Army Corps of Engineers, Naval Facilities Engineering Command, Air Force Civil Engineer Support Agency. (2008). Technical Manuals, Unified Facilities Criteria (UFC), Structures to resist the effects of accidental explosions, UFC 3-340-02.

Zhdan, S. A. (1983). Calculation of gas mixture explosion with regard to shift of chemical equilibrium products. *Journal of Physics of Combustion and Explosion (Russian Academy of Sciences) 1*, 131.

Chapter 7
Blast Wave Experiments of High Explosives

Isabelle Sochet

Applications of pyrotechnic materials cover an extensive area: protection of the life and security of the population, satellites, tactical and ballistic missiles, space launchers, aeronautics industry, automotive security (airbags), rail signal devices, perforation charges for petroleum industries, demolition in mines, quarries, buildings, etc. and a large variety of bombs for terrorist attacks.

There is a large variety of high explosives which exist on various forms. The use of trinitrotoluene (TNT) as a reference explosive in properties of blast is universal. Hence, it is very important for experiments to have an absolutely reproducible reference. The difficulty is in the use of small charges. The initiator system and the booster occupy a nonnegligible mass. Hence, the expansion of detonation products and the driving mechanism for the blast wave may be affected.

The literature of blast properties is well documented (see References) and the analysis conducted here is focused on small charges of TNT, C4 and composition B from experiments realised by the French-German Research Institute of Saint-Louis (ISL) and the French Alternative Energies and Atomic Energy Commission (CEA, Gramat). The properties of blast wave are analysed in terms of overpressure, arrival time, impulse, duration and compared with the abacus of UFC-3-340-2 (U.S. Army Corps of Engineers 2008). Fitted laws are given to calculate the overpressure and arrival time versus scaled distance expressed in terms of mass. The reverse laws are equally established.

I. Sochet (✉)
Laboratoire PRISME, INSA Centre Val de Loire, Bourges, France
e-mail: isabelle.sochet@insa-cvl.fr

I. Sochet (ed.), *Blast Effects*, Shock Wave and High Pressure Phenomena,
https://doi.org/10.1007/978-3-319-70831-7_7

7.1 Experimental Setup

Two experimental setup are considered here. The first one is devoted to spherical high explosive (HE) and the second to hemispherical charges.

The experiments realised at CEA (The French Alternative Energies and Atomic Energy Commission) concern only hemispherical charge detonated on the ground. An ideal surface reflection is obtained with a reinforced concrete platform dedicated for the hemisphere experiments (Lefrançois et al. 2002). A TNT charge of 22.8 kg is tested. The initiation train occurs with a detonator, a detonating cord and a small cylindrical booster and a hemispherical booster (250 g weight and a HMX/Viton 95/5 composition). A shock absorber media allows to reduce the pressure sensor vibration and split off the measurement platform from the hemispherical charge support. The pressure gauges are PCB piezoelectric gauges with integrated amplifier, the signals are recorded by a Nicolet acquisition card system. The pressure gauges are distributed at several distances from the centre of charge: 2, 3, 4, 6, 10, 12, 14 and 15 m.

The experiments conducted at ISL (French-German Research Institute of Saint-Louis) concern hemispherical and spherical charges. ISL hemispherical concrete blast pad is presented in Fig. 7.1.

Explosive charges are placed at the centre of the metallic protection plate of the concrete slab, at ground level. Blast propagation is recorded by 5 successive side-on pressure gauges located at 3, 5, 7, 10 and 15 m from the charge and at 1.5 m from the slab surface. Two types of explosives are considered: TNT and C4. The TNT charges are cylindrical with a mass of 1 kg (length 0.783 m, diameter 0.10 m, $L/D = 7.83$) and 1.8 kg (length 0.9243 m, diameter 0.13 m, $L/D = 7.11$). The booster is a charge

Fig. 7.1 Hemispherical blast pad

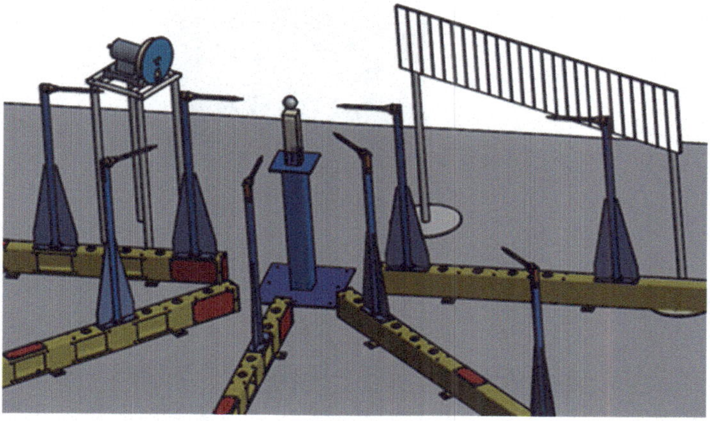

Fig. 7.2 Spherical blast pad

of C4 of 0.055 and 0.110 kg, respectively. The charge of C4 is a molded hemisphere of 2 kg. The cylindrical charges are positioned vertically and the ignitor at the top of the charge.

Figure 7.2 shows the experimental setup designed for the study of spherical blast waves. An explosive charge is placed 1.5 m above the concrete test pad (i.e. HOB), on a polyurethane block of foam fixed on the top of the central metallic pillar. This configuration helps delaying the interference of the reflected leading shock on the ground surface. The evolution of the pressure around the charge is recorded by 8 PCB (137A22) piezoelectric side-on pressure sensors mounted on fixed poles located, respectively, at 0.5 (2 sensors), 0.75, 1, 1.25, 1.5, 2 and 3 m from the charge centre. Sensor positions were chosen to limit their wake influence on each other. The first two sensors placed on both sides of the charge at 0.5 m are used to verify the spherical geometry of the explosion.

A spherical 0.031 kg cast Composition B charge (Fig. 7.3) was specifically produced to calibrate the experimental setup. Ignition was realised by an RP83 high voltage detonator surrounded by 0.003 g of C4 as a relay.

Blast effects of charges of molded Composition B are analysed. For that, two masses are tested. The first one is 0.036 kg with 0.004 kg of C4 for the booster and 0.001 kg for the detonator and the second one is 0.063 kg with 0.005 kg of C4 for the booster and 0.001 kg for the detonator. Spherical charges of TNT charges are also used with the booster C4. The masses are 0.941 kg including 0.050 kg of C4, 1.899 kg with 0.120 kg of C4 and 1.901 kg including 0.120 kg of C4.

Fig. 7.3 Spherical 31 g Comp-B charge

7.2 Blast Characteristics

Numerous experiments have established relationships to describe the air-blast parameters from incident and reflected shocks. The analysis of blast wave propagation is characterised by the arrival time, overpressure and impulse. Empirical laws are deduced for each detonation explosive charges. In this section, the hemispherical surface bursts and free-air are examined without reflection on the ground or any structure.

The classical references are Brode (1955), Held et al. (1961), Adushkin and Korokov (1961), Baker (1973), Henrich (1979), Kinney and Graham (1985), Mills (1987) and Gelfand (2004). These various approaches are not reported here. Today, the main reference is Unified Facilities Criteria (U.S. Army Corps of Engineers 2008). This technical report presents a compilation of blast parameters generated by several types of explosives in different geometries. The results are included in CONWEP (Protective Design Center 2007) which is a collection of conventional weapons effects calculations. Kingery and Bulmash (1984) have formulated the characteristics of blast from spherical and hemispherical of 1 kg of TNT. Swisdak (1994) has revised Kingery the equations for a hemispherical charge of TNT and gives expressions of arrival time, incident and reflected overpressure, incident and reflected positive impulse versus the scaled distance Z. More recently, Shin et al. (2015) have proposed modified charts and polynomials for air-blast parameters using 1D and 2D simulations with Autodyn. Hence, the reference considered here is Unified Facilities Criteria widely used by engineers and researchers community.

The incident overpressure in Fig. 7.4 illustrates the sensitivity of the geometry of the charge of TNT. The experiments with the highest spherical charge are in a perfect agreement with the reference curve (UFC-3-340-2 (U.S. Army Corps

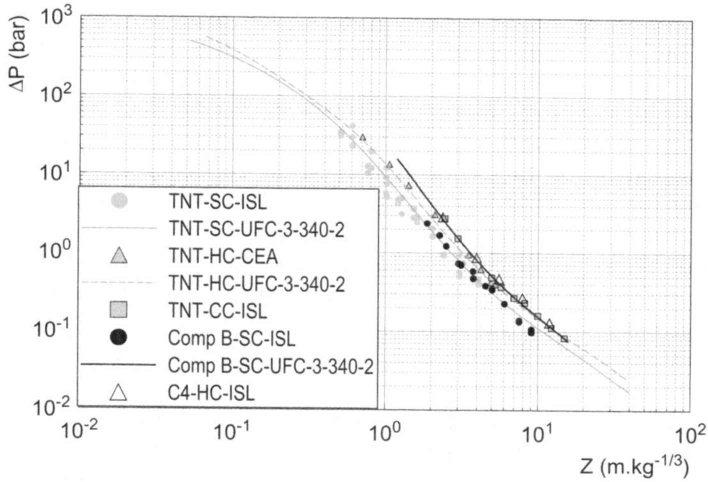

Fig. 7.4 Overpressure versus scaled distance in mass for high explosives

of Engineers 2008)). The very elongated effect of the cylindrical explosive (ratio of length/diameter in order of 7) is pronounced in near field. The overpressure is more important for the cylindrical charge that the spherical or hemispherical one for the same scaled distance. The gap decreases with the distance and a convergence with the reference of UFC for hemispherical charge is reached at $Z = 5\,\mathrm{m\,kg^{-1/3}}$. Hence, it can be explained by predominant direction of resulting detonation products due to the shape of the charge. The overpressure obtained with a charge of C4 converges towards the reference of TNT with increasing of scaled distance. The effect of geometry of the charge on the overpressure is not reproduced on the reduced arrival time. The effect of the geometry of the charge is highlighted by Peugeot et al. (2006). The blast waves generated by the detonation of spherical and cylindrical high explosives do not present the same characteristics. Thus, in case of a cylindrical charge in near field the shock is neither spherical nor cylindrical and gradually becomes spherical at far field. The authors show that for scaled distances $Z < 2\,\mathrm{m\ kg^{-1/3}}$, the overpressure is much higher with a ratio L/D (Length/Diameter of charge) of 4 than for a ratio of $1/4$. In first case, the explosive energy release is mainly radial in opposite to the second case which has a lateral surface too small and so the overpressure is smaller. For scaled distances $Z > 2\,\mathrm{m\ kg^{-1/3}}$, the geometry is not significant. As a result, the small scale test realised with a cylindrical charge of TNT ($L/D = 7$) is in total agreement with the analysis of Peugeot et al. (2006).

The analysis of blast parameters generated by spherical charges shows the impact of composition of explosive and the effect of scale of trials. The charge of TNT conducts to an evolution of overpressure slightly parallel to the reference while the overpressure from Composition B charge diverges with the increasing scaled distance. Comparing results of the Composition B with the reference (dashed line Fig. 7.4) shows that the size of explosive charge has a major effect. The presence of the booster and the detonator represents 14% of total mass for the charge 0.036 kg and 9.5% for the another charge (0.063 kg). The percentage is very higher than for charges of several kilograms. Generally the relay is in order of 0.030 kg for large masses of explosive, meaning 3% for 1 kg or 0.6% for 5 kg of explosive. Consequently, for small charges, the relay impacts the energy of explosion and the strength of shock wave. And so the blast effect of a small charge of Composition B is quite similar to a TNT charge. As regards the experiments with TNT charge, the small masses of charges lower than 2 kg indicate a global under estimation of the overpressure and a discrepancy of values in near field. Hence, it means that the considered masses are not representative of the reference UFC-3-340-2 (U.S. Army Corps of Engineers 2008). This behaviour is noted by Dewey (1964) and Kleine et al. (2003). The authors point out that TNT cannot be reliably detonated in amounts less than 4 kg due to the presence of booster.

Considering the arrival time evolution in function of scaled distance a good correlation is obtained for C4 and TNT charges in case of spherical and hemispherical charges (Fig. 7.5). However, in near field ($Z < 1\,\mathrm{m\,kg^{-1/3}}$) the arrival times from TNT are not correlated with the reference UFC-3-340-2 (U.S. Army Corps of Engineers 2008). The convergence is obtained from $Z = 2\ \mathrm{m\,kg^{-1/3}}$. The arrival time for the composition B is consistent with the abacus of TNT (UFC-3-340-2 (U.S. Army Corps of Engineers 2008)) (Fig. 7.5).

The evolution of positive impulse versus scaled distance shows clearly the disparity of experimental results for the spherical charges of TNT and Composition B (Fig. 7.6). They include a type of "point cloud". Hence, the experimental results of positive impulse for these explosives at small scale cannot prove the evolution given by UFC-3340-2 (U.S. Army Corps of Engineers 2008). Nevertheless, the positive impulse obtained with the cylindrical charge of TNT is well organised. They are

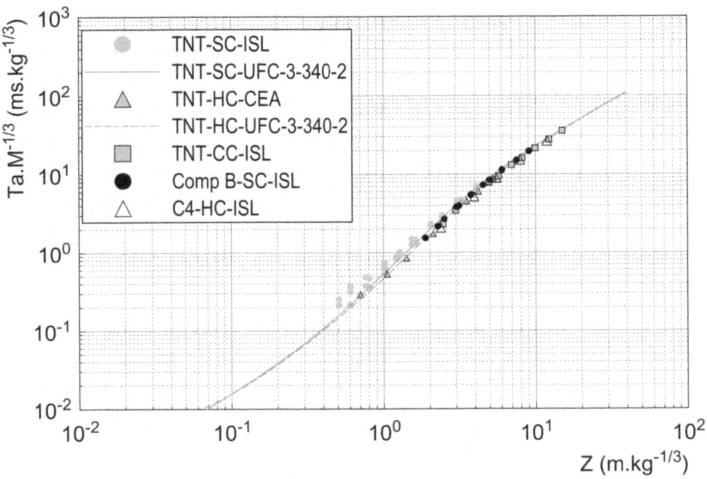

Fig. 7.5 Reduced arrival time versus scaled distance in mass for high explosives

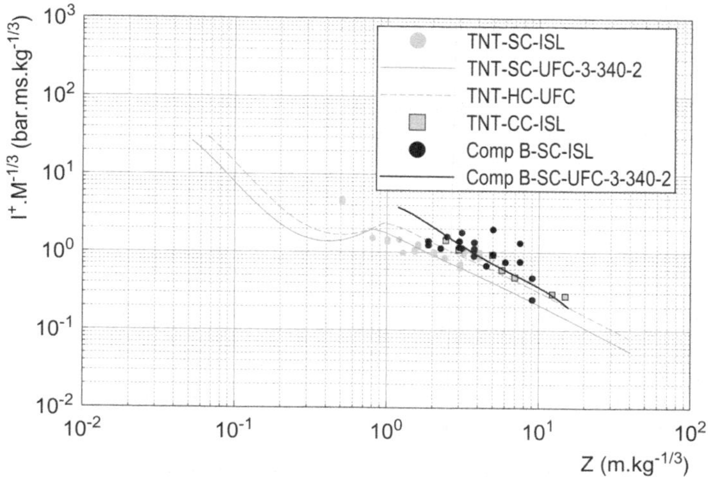

Fig. 7.6 Reduced positive impulse versus scaled distance in mass for high explosives

close to the abacus of hemispherical charge from UFC-3-340-2 (U.S. Army Corps of Engineers 2008). The determination of positive impulse is particularly difficult partly due to the lack of sharpness of the signal, the response of the pressure transducer. These two factors may contribute to strong discrepancy of the results from the reference UFC-3-340-2 (U.S. Army Corps of Engineers 2008).

This analysis indicates that the hemispherical explosive of C4 and cylindrical charge of TNT led to more important overpressure peak and positive impulse than TNT charge for a similar arrival time at given scaled distance.

Note 1 In Appendix C of this book the reader will find all polynoms laws to calculate the overpressure and arrival time versus scaled distance expressed in terms of mass. The reverse laws are equally reported.

Acknowledgements I gratefully acknowledge A. Lefrançois from CEA (French Alternative Energies and Atomic Energy Commission) and M.O. Sturtzer from ISL (French-German Research Institute of Saint-Louis) for transmitting experimental data and for discussions.

References

Adushkin, V. V., & Korokov, A. I. (1961). Parameters of a shock wave near to HE charge at explosion in air. *Prikladnoi Mekhaniki i Tekhnicheskoi Fiziki, 5*, 119–123 (in Russian).

Baker, W. E. (1973). *Explosions in air*. Austin: University of Texas Press.

Brode, H. L. (1955). Numerical solutions of spherical blast waves. *Journal of Applied Physics, 26*(6), 766.

Dewey, J. M. (1964). The air velocity in blast waves from T.N.T. explosions. *Proceedings of the Royal Society A, 279*, 366–385.

Gelfand, B. (2004, April). Translation from Russian to English The Book "Blast effects caused by explosions" authored by B. Gelfand and M. Silnikov. United States Army, European Research Office of the U.S. Army, London, England. Contract Number N62558-04-M-0004.

Held, M., Jager, E. H., & Stolzl, B. (1961). *TNT-blast equivalence forbursting or pressurized gas conventional vessels*. Paper at 6th SMIRT conference, Paris.

Henrich, J. (1979). *The dynamics of explosions*. Elsevier: Amsterdam.

Kingery, C. N., & Bulmash, G. (1984, April). *Airblast parameters from TNT spherical air burst and hemispherical surface burst*. US Technical Report ARBRL-TR-02555. Aberdeen Proving Ground, MD: Ballistics Research Laboratory.

Kinney, G. F., & Graham, K. J. (1985). *Explosives shocks in air* (2nd ed.). Berlin: Springer.

Kleine, H., Dewey, J. M., Oashi, K., Mizuka, T., & Takayama, K. (2003). Studies of the TNT equivalence of silver azide charges. *Shock Waves, 13*, 123–138.

Lefrançois, A., Baudin, G., Cremoux, J. L., Massoni, J., & Saurel, R. (2002). Blast efficiency of aluminized high explosives. In *17th International Symposium on the Military Application of Blast and Shock (MABS17)*, Las Vegas, NV.

Mills, C. A. (1987). The design of concrete structure to resist explosions and weapon effects. In *Proceedings of the 1st International Conference on Concrete for Hazard Protections*, Edinburgh, UK (pp. 61–73).

Peugeot, F., Deschalbault, E., & Péron, P. F. (2006, October). TNT equivalency: misconceptions and reality. Munitions Safety Information Analysis Center, MSIAC Unclassified, L-132.

Protective Design Center, United States Army Corps of Engineers. (2007). CONWEP, Conventional Weapons Effects. Retrieved May 22, 2007. https://pdc.usace.army.mil/software/conwep/

Shin, J., Whittaker, A., Cormie, D., & Willford, M. (2015). Design charts and polynomials for air-blast parameters. In M. G. Stewart & M. D. Netherton (Eds.), *Third International Conference on Protective Structures (ICPS3)*, Newcastle, 3–6 February 2015.

Swisdak, M. M. J. (1994). Simplified Kingery airblast calculations. In *Proceedings of the 26th Department of Defense (DoD) Explosives Safety Seminar*, Miami, FL.

U.S. Army Corps of Engineers, Naval Facilities Engineering Command, Air Force Civil Engineer Support Agency. (2008). Technical Manuals, Unified Facilities Criteria (UFC), Structures to resist the effects of accidental explosions, UFC 3-340-02.

Chapter 8
TNT Equivalency

Isabelle Sochet

The energy released by the detonation of fuel-air (or fuel-oxygen) or high explosive has a significant effect on the blast wave propagation independently of the environment. The unconfined or confined environment with or without obstructions affects the propagation of the initial shock through the creation of multiple reflection shocks. The rate of energy is a very sensitive quantity that governs the combustion, the expansion of detonation products, and the driving mechanism for the blast wave. The energy released depends on the chemical composition of the explosive itself. The universal TNT explosive implies that it is convenient to compare the effects of fuel-air or high explosive detonation to TNT. This comparison can be conducted using the TNT equivalency. The concept of TNT equivalency is given by a proportional ratio of the TNT mass to explosive mass, which produces a blast wave of equal magnitude.

However, the time histories of the blast waves from any two different explosives are never exactly the same. If there is a match of peak pressures from two different explosives, the wave profiles, the positive durations, the positive phase impulses or the time-of-arrival of the second shocks should not also be expected to match. Any of these measures can be used to determine a TNT equivalence and lead to a different equivalence for each measure. The two measures most commonly compared to obtain TNT equivalence are peak hydrostatic overpressure and hydrostatic pressure impulse, and these give different values of equivalence. It is important, therefore, to state which parameter is being compared, to obtain the equivalence value and its variation with distance.

I. Sochet (✉)
Laboratoire PRISME, INSA Centre Val de Loire, Bourges, France
e-mail: isabelle.sochet@insa-cvl.fr

I. Sochet (ed.), *Blast Effects*, Shock Wave and High Pressure Phenomena,
https://doi.org/10.1007/978-3-319-70831-7_8

Here, thermodynamics of explosions and the difficulty in adopting a universal value for the detonation energy are exposed and illustrated through TNT and composition B. Some definitions of TNT equivalence are given and applications to the blast properties of explosives studied in the previous chapter are discussed.

8.1 Thermodynamics of Explosions

The thermodynamics of an explosion is an important point in determining the products of explosives as a function of the decomposition reactions and the energy released by the reactions. The energy or heat released by the chemical reaction that occurs during the combustion of a propellant or detonation of an explosive is called the 1"heat of explosion" or "heat of detonation." This value describes only the heat caused by the reaction of the explosive itself yielding the detonation products; therefore, this value does not include any heat generated by secondary reactions of the explosive or its products with air. Generally, the term "heat of explosion" is used for propellants, and the term "heat of detonation" is used for explosives. The combustion energy is a special case of reaction energy, it is the reaction energy required for the combustion of a reactant with a given amount of oxygen such that it is completely oxidized.

In an actual explosion, the composition of the products is not always the same for a given explosive. Factors such as the initial temperature and density, degree of confinement, particle size and morphology, and the size and shape of the load affect the pressure and temperature behind the detonation front, where the products are rapidly expanding, and a balance between the products is not achieved.

There is some confusion regarding the terminology related to the energy and heat of detonation, and the two are often used interchangeably. The heat of detonation is determined in a closed calorimetric tank and does not account for the energy available from the highly compressed gas products, which contribute to the amount of energy transmitted to the blast wave, as outlined by Scilly (1995).

Thus, the term "detonation energy" is used hereafter to refer to the calculated detonation energy of an explosive without the presence of air.

8.1.1 Oxygen Balance

The detonation of an explosive is an oxidation reaction for which it can be assumed that all of the carbon forms CO_2, all of the hydrogen forms water, and all of the nitrogen forms N_2. For the explosive composition $C_aH_bO_cN_dCl_eS_f$, the oxidation reaction is represented by:

$$C_aH_bO_cN_dCl_eS_f \rightarrow a\,CO_2 + 0.5\,(b-e)\,H_2O + 0.5\,d\,N_2 + e\,HCl + f\,SO_2 + [0.25\,(e-b) - (a+f) + 0.5\,c]\,O_2 \quad (8.1)$$

The concentration of oxygen atoms in an oxidant is given by an oxygen balance (OB). This is an important term that indicates the oxidation potential and the number of molecules of oxygen remaining after the oxidation of H, C, Mg, Al, etc., to produce H_2O, CO_2, MgO_2. If the amount of oxygen in an explosive is limited and insufficient to obtain a complete oxidation reaction, then the amount of oxygen that is required to complete the reaction is preceded by a negative sign and the OB is negative. The oxygen balance is expressed as a percentage by mass:

$$OB = \frac{[0.25\,(e-b)-(a+f)+0.5\,c]\times 32}{MW_{explosive}}\times 100\,(\%) \tag{8.2}$$

where $MW_{explosive}$ is the molecular mass of the explosive and is expressed in $g\,mol^{-1}$. In conclusion, the OB provides information about the products: a positive value indicates an excess of oxygen in the explosive, whereas a negative value indicates that oxygen must be provided, typically by the surrounding air. If the OB is mostly negative, then there is not enough oxygen to form CO_2 and thus toxic gases such as CO form. In the case of negative OB, the detonation products can also react with oxygen from the air in a postcombustion process and lead to larger blast effects. Regarding mixtures of explosives, detonation depends not only on the OB, but also on combustion reactions and other physical properties.

The OB provides no information on the exchange of energy during the explosion. For 2,4,6-Trinitrotoluene (TNT), which has the chemical formula $C_7H_5N_3O_6$ and a molecular weight of 227 $g\,mol^{-1}$, the OB is mostly negative (−74%).

8.1.2 Decomposition Rules

To clarify the problem related to the decomposition products (Akhavan 2004), a set of rules was developed. The rules of decomposition are known as the Kistiakowsky–Wilson (KW) rules, which are used for explosives in which oxygen deficits are moderate and the OB is higher than −40%. They are given as follows:

- Rule 1: The carbon atoms are converted into CO
- Rule 2: If oxygen remains, then hydrogen is oxidized to water
- Rule 3: If oxygen still remains, then CO is oxidized to CO_2
- Rule 4: All nitrogen is converted into N_2 gas

For explosives with an OB lower than −40%, the modified KW rules are as follows:

- Rule 1: The hydrogen atoms are converted into water
- Rule 2: If oxygen remains, then the carbon is converted into CO
- Rule 3: If oxygen still remains, then CO is oxidized to CO_2
- Rule 4: All nitrogen is converted into N_2 gas

For TNT ($C_7H_5N_3O_6$), the products resulting from each rule are as follows:

- Rule 1: 5 H $\rightarrow$ 2.5 H_2
- Rule 2: The remaining atoms of oxygen 6 – 2.5 = 3.5 gives 3.5 O $\rightarrow$ 3.5 CO
- Rule 3: There is no more oxygen
- Rule 4: 3 N $\rightarrow$ 1.5 N_2

The reaction of decomposition of TNT is:

$$C_7H_5N_3O_6 \rightarrow 3.5CO + 3.5C + 2.5H_2O + 1.5N_2 \tag{8.3}$$

The Springall–Roberts rules (SR rules) are based on the KW rules (unmodified) and apply two additional conditions:

- Rule 1: The carbon atoms are converted into CO
- Rule 2: If oxygen remains, then hydrogen is oxidized to water
- Rule 3: If oxygen still remains, then CO is oxidized to CO_2
- Rule 4: All nitrogen is converted into N_2 gas
- Rule 5: One-third of the CO formed is converted into carbon and CO_2
- Rule 6: A sixth of the original value of CO is converted with hydrogen to form C and water

The SR rules for TNT ($C_7H_5N_3O_6$) lead to the following decomposition:

- Rule 1: 6 C $\rightarrow$ 6 CO
- Rule 2: There is no more oxygen
- Rule 3: There is no more oxygen
- Rule 4: 3 N $\rightarrow$ 1.5 N_2
- Rule 5: 1/3 (6 CO) $\equiv$ 2 CO $\rightarrow$ C + CO_2
- Rule 6: 1/6 (6 CO) $\equiv$ 1 CO and 1 CO + H_2 $\rightarrow$ C + H_2O

That is: $C_7H_5N_3O_6 \rightarrow$ 6 CO + C + 1.5 N_2 + 2.5 H_2, and applying rules 5 and 6, the products are:

$C_7H_5N_3O_6 \rightarrow (C + CO_2 + C + H_2O) + 3\ CO + C + 1.5\ N_2 + 1.5\ H_2$

Ultimately, the overall reaction is:

$$C_7H_5N_3O_6 \rightarrow 3C + 3CO + CO_2 + H_2O + 1.5H_2 + 1.5N_2 \tag{8.4}$$

According to Scilly (1995), the decomposition equation recommended by Kamlet–Jacob (KJ rule) can be obtained from:

$$C_aH_bO_cN_d \rightarrow (a-0.5\,d+0.25\,b)\,C+(0.5\,d-0.25\,b)\,CO_2+0.5\,b\,H_2O+0.5\,c\,N_2$$

In this scheme, CO is not preferentially formed, and CO_2 is assumed to form as the only oxidation product of carbon and water and always forms at the beginning of the reaction. For TNT, the decomposition is then:

$$C_7H_5N_3O_6 \rightarrow 5.25C + 1.75CO_2 + 2.5H_2O + 1.5H_2 + 1.5N_2 \tag{8.5}$$

Kinney and Graham (1985) considered that all of the oxygen is contained in the carbon monoxide, which implies, in the case of TNT, the following chemical equation:

$$C_7H_5N_3O_6 \rightarrow C + 6CO + 2.5H_2 + 1.5N_2 \quad (8.6)$$

8.1.3 Detonation Energy

The detonation energy can be calculated using the Helmholtz free energy method and the enthalpy method. The detonation energy value is not accurate. In fact, it depends on the products considered, the heat of formation for the different species available in the literature, and whether or not it is an experimental value. Hence, in a numerical simulation, the value of the detonation energy can have an impact on the expansion of the compressed product gases and the interface between these products and shocked air, and consequently the near field blast.

8.1.3.1 Helmholtz Free Energy Method

The detonation energy corresponds to the explosive energy transmitted from the shock wave and the associated wind, i.e., the work done in the expansion of gases produced during the explosion, and is given by $\int_{\text{Initial}}^{\text{Final}} P.\Delta V$. By applying the first and second laws of thermodynamics, the change in the Helmholtz free energy can be used to calculate the energy of the explosion expressed in terms of the internal energy ΔU and entropy ΔS: $\Delta F = \Delta U - T\Delta S$ with $\Delta U = \Delta H - RT\Delta N$ and $\Delta H_{\text{detonation}} = \Delta H_P - \Delta H_R$; $\Delta n = n_P - n_R$; $\Delta S = \Delta S_P - \Delta S_R$. The subscripts "*P*" represent the products, whereas "*R*" represent the reactants.

8.1.3.2 Enthalpy Method

The reaction energy is the energy released by the reaction, which is calculated by the enthalpy change involved in the chemical reaction between the standard state reaction products and the reactants (i.e., enthalpy of formation at normal pressure and temperature conditions): $\Delta H_{\text{detonation}} = \Delta H_P - \Delta H_R$.

8.1.4 Application to TNT

Tongchang et al. (1995) performed experiments using a blast calorimeter in which the bomb had a cylindrical internal volume of 5 L and could support a pressure of 200 MPa. The experiments were performed with maximum loads of

50 g. The explosive force was measured according to the nature of the cartridge (porcelain, brass) and its thickness. All of the tests resulted in values between −4.31 and − 4.40 MJ kg^{-1}. These values are in agreement with the values used by Gelfand (2004), −4.517 MJ kg^{-1}, Baker (1973), −4.520 MJ kg^{-1}, Cooper (1996) −4.56 MJ kg^{-1}, Pförtner (1977), −4.686 MJ kg^{-1}. Other values for the explosive force are cited by Filler (1956), including the calculated energy in the Encyclopedia of Chemical Technology, −3.87 MJ kg^{-1}, the results of Tonegutti using a 2 g charge with a conventional calorimeter (detonation energy of −3.210 MJ kg^{-1}) and, for a load of 100 g, the energy measured by the Armament Research Establishment using calorimetry (−4.535 MJ kg^{-1}). Omang et al. (2009) used an energy value of −4.260 MJ kg^{-1} to characterize the propagation of shock waves following the detonation of spherical and hemispherical loads of 1 kg of TNT. Scilly (1995) reported experimental values of −4.27 MJ kg^{-1} for water in a liquid state and −4.573 MJ kg^{-1} with water vapor, and calculated values of −4.608 MJ kg^{-1} with liquid water and −4.918 MJ kg^{-1} with water vapor. The gaseous equation of state (EOS) affects the detonation energy. In Cheetah code, the application of BKWC and BKWS EOS lead a detonation energy of −4.91 and −4.75 MJ kg^{-1} respectively (Private communication with Lapebie E. from the CEA, French Alternative Energies and Atomic Energy Commission).

Thus, all of the detonation energies reported here range from a minimum of −4.918 MJ kg^{-1} (Scilly 1995) to a maximum of −3.210 MJ kg^{-1} (Filler 1956). The average value is $|\Delta H_{\mathrm{detonation}}| = 4.43\ \mathrm{MJ\,kg^{-1}}$.

In the following applications, thermochemical data are extracted from Meyer et al. (2007) and reported in Tables 8.1 and 8.2.

Knowing the standard enthalpies of formation of reaction products and reactants or components, the detonation energy is calculated for each decomposition developed above.

$$\Delta H_{\mathrm{detonation}} = \sum \Delta H^0_{f,\,\mathrm{products}} - \sum \Delta H^0_{f,\,\mathrm{reactants}} \tag{8.7}$$

Table 8.1 Enthalpy of the formation of gaseous reaction products (Meyer et al. 2007)

		Molecular weight (Mw)	Standard enthalpy of formation ΔH_f^0
Products	Formula	g mol^{-1}	kJ mol^{-1}
Carbon monoxide	CO	28.01	−110.6
Carbon dioxide	CO_2	44.01	−393.8
Water vapor	H_2O	18.02	−242.0
Water liquid	H_2O	18.02	−286.1

Table 8.2 Enthalpy of formation of TNT—Meyer et al. (2007)

		Molecular weight (Mw)	Standard enthalpy of formation ΔH_f^0	
Component	Formula	g mol^{-1}	kJ kg^{-1}	kJ mol^{-1}
TNT (Trinitrotoluene)	$C_7H_5N_3O_6$	227.1	−295.3	−67.06

The results are:

1. In considering water vapor
   (a) Modified KW rules (8.3): $\Delta H_{\text{detonation}} = -4.07\,\text{MJ kg}^{-1}$
   (b) SR rules (8.4): $\Delta H_{\text{detonation}} = -3.96\,\text{MJ kg}^{-1}$
   (c) KJ rules (8.5): $\Delta H_{\text{detonation}} = -5.40\,\text{MJ kg}^{-1}$
   (d) Kinney rule (8.6): $\Delta H_{\text{detonation}} = -2.63\,\text{MJ kg}^{-1}$
2. In considering liquid water
   (a) Modified KW rules (8.3): $\Delta H_{\text{detonation}} = -4.56\,\text{MJ kg}^{-1}$
   (b) SR rules (8.4): $\Delta H_{\text{detonation}} = -4.16\,\text{MJ kg}^{-1}$
   (c) KJ rule (8.5): $\Delta H_{\text{detonation}} = -5.89\,\text{MJ kg}^{-1}$

The application of the Kinney rule (8.6) appears to be inappropriate for calculating the detonation energy. The scheme gives too low a value equal to $-2.63\,\text{MJ kg}^{-1}$ and independent of the water state. In contrast, the KJ decomposition yields high values.

As expected, higher values are obtained with liquid water as a result of the standard enthalpy of formation.

Consequently, it can be stated that modified KW decomposition gives an adequate value of detonation energy ($\Delta H_{\text{detonation}} = -4.56\,\text{MJ kg}^{-1}$) considering water vapor and is in good agreement with the results of the literature review.

8.1.5 Application to Composition B

The Composition B used by the French–German Research Institute of Saint-Louis (ISL) is a mixed explosive composition of 65% RDX and 35% TNT in terms of mass. The element formula of RDX is $C_3H_6O_6N_6$ and the thermochemical data are given in Table 8.3.

The composition B is defined by $0.65\ C_3H_6O_6N_6 + 0.35\ C_7H_5N_3O_6$ for 1 kg of mixture. The OB for a mixture is proportional to each component and is given by:

$$\text{OB}_{\text{mixture}} = \sum x_i \text{OB}_i \qquad (8.8)$$

where x_i is the molar proportion of the component i characterized by the OB_i.

Thus, $x_{\text{RDX}} = 2.927$ and $x_{\text{TNT}} = 1.541$ and the OB for composition B is equal to: −39.94%. It is the limit between the application of the KW rules and

Table 8.3 Enthalpy of formation of RDX—Meyer et al. (2007)

		Molecular weight (Mw)	Standard enthalpy of formation ΔH_f^0		Oxygen balance (OB)
Component	Formula	g mol^{-1}	kJ kg^{-1}	kJ mol^{-1}	%
Hexogen (RDX)	$C_3H_6O_6N_6$	222.1	+301.4	+66.87	−21.61

the modified KW rules, as defined above. To define the decomposition reaction, the atomic composition of the mixture is first determined by means of the molar proportion, which yields to: $C_{19.568}H_{25.267}O_{22.185}N_{26.808}$.

The application of unmodified KW rules leads to:

$$C_{19.568}H_{25.267}O_{22.185}N_{26.808} \rightarrow 19.568CO + 7.24H_2O + 5.3935H_2 + 11.0925N_2 \quad (8.9)$$

If the modified KW rules are considered because of the limit of OB, then the decomposition is as follows:

$$\begin{aligned} C_{19.568}H_{25.267}O_{22.185}N_{26.808} \rightarrow\ & 14.1745CO + 5.3935C \\ & + 12.6335H_2O + 11.0925N_2 \end{aligned} \quad (8.10)$$

According the rules of SR, the decomposition is:

$$\begin{aligned} C_{19.568}H_{25.267}O_{22.185}N_{26.808} \rightarrow\ & 3.26135CO_2 + 9.78397CO + 6.52268C \\ & + 10.50133H_2O + 2.13217H_2 \\ & + 11.0925N_2 \end{aligned} \quad (8.11)$$

The detonation energy may be calculated for each decomposition model by applying the general equation (8.7) to composition B with thermochemical data in Tables 8.1 and 8.3. The enthalpy of composition B is proportional to each component and equals to +92.39 kJ. The results are:

1. In considering water vapor
   (a) KW rule (8.9): $\Delta H_{detonation} = -4.00\,\text{MJ kg}^{-1}$
   (b) Modified KW rule (8.10): $\Delta H_{detonation} = -4.72\,\text{MJ kg}^{-1}$
   (c) SR rule (8.11): $\Delta H_{detonation} = -4.99\,\text{MJ kg}^{-1}$
2. In considering liquid water
   (a) KW rule (8.9): $\Delta H_{detonation} = -4.33\,\text{MJ kg}^{-1}$
   (b) Modified KW rule (8.10): $\Delta H_{detonation} = -5.27\,\text{MJ kg}^{-1}$
   (c) SR rule (8.11): $\Delta H_{detonation} = -5.46\,\text{MJ kg}^{-1}$

As expected, the detonation energy calculated by Akhavan (2004) for a slightly different composition (60% RDX + 40% TNT) is $-4.72\,\text{MJ kg}^{-1}$.The calculation is conducted using the modified KW rule and by considering water in its vapor state. Tongchang et al. (1995) has investigated the influence of the ratio of TNT/RDX on the detonation energy. The experimental approach is conducted using an adiabatic detonation calorimeter for simple high explosives and a linear evolution between the detonation energy of the explosives and the proportion of TNT/RDX has been defined. Thus, for a ratio TNT/RDX of 35/65 the detonation energy is -5.37 and $-5.26\,\text{MJ kg}^{-1}$ for a ratio 40/60. This result is in agreement with the above calculated value using the modified KW rule when considering liquid water. According to UFC-3-340-2 (U.S. Army Corps of Engineers 2008) the energy is $-6.43\,\mathrm{MJ\,kg^{-1}}$without specification of the proportion RDX/TNT and the way in which the value is determined.

The analysis conducted here for two high explosives, TNT and composition B, demonstrates the sensitivity of the decomposition scheme, the gaseous equation of state, thermochemical data, and the state of water regarding the value of the detonation energy. It explains the panel of available values in the free literature and at same time the difficulty in adopting the suitable value.

The energy from the explosion released by the chemical reaction of fuel in the air or oxygen is to the same order of magnitude or higher than the detonation energy of high explosives. The energies calculated earlier for gaseous explosions (Chap. 6) are on average $3\,\mathrm{MJ\,kg^{-1}}$for mixtures of ethylene–air and propane–air (experiments of Fraunhoffer, Brossard and Dorofeev) and to the order of 8–10 $\mathrm{MJ\,kg^{-1}}$ for mixtures of propane–oxygen (experiments of Dewey and Prisme). Hence, to compare the properties of blast waves, one solution consists in determining TNT equivalency for fuel and high explosives.

8.2 TNT Equivalency

8.2.1 Overview

Generally, the TNT equivalency represents the mass of TNT that would result in an explosion of the same energy level as the unit weight of the explosive under consideration. Specifically, TNT equivalent is defined as the ratio of the mass of TNT to the mass of the explosive that results in the same magnitude of blast wave (or impulse pressure) at the same radial distance for each charge, which assumes the scaling laws of Sachs and Hopkinson. All explosives generate blast waves that exhibit similar characteristics. The primary reason for choosing TNT as the reference explosive is that there is a large amount of experimental data regarding the characteristics of blast waves associated with this explosive. There are several methods for determining the explosive characteristics of different explosives, but they do not yield the same values as for the TNT equivalent. These values depend on the characteristic parameters of the blast wave, the geometry of the load, and the distance from the explosive charge. The mechanism of energy release during the detonation process varies depending on the nature of the explosive. Explosives are generally composed of two parts: an oxidizer and a fuel. Pure explosives, which are called "ideal" explosives, have a threshold of molecules for each component defined for different mixtures of explosives. Because of the effects of the shock wave during detonation, the oxidizer and fuel interact in the area near the chemical reactions. The speed of the chemical reaction and the detonation velocity for ideal explosives are greater because more favorable conditions result in a greater efficiency. However, a larger detonation velocity results in a higher rate of energy release. Furthermore, the amount of energy released is directly proportional to the overpressure and the impulse of the blast wave. The reaction between gaseous products due to degradation of the explosive for a non-ideal explosive occurs beyond the chemical reaction zone, and the energy released after the reaction cannot sustain the blast wave.

In summary, an overview of different approaches to the determination of a TNT equivalent is given. It is possible to distinguish between approaches based on the pressure, impulse, Chapman–Jouguet state, and explosion yield.

8.2.2 *Definitions*

8.2.2.1 Pressure-Based Concept

Esparza (1986) based the equivalence of the incident pressure on the mass ratio of TNT (M_{TNT}) to the considered explosives (M) that produces the same peak hydrostatic overpressure at the same radial distance of each load. The equivalent mass of an explosive pressure is then:

$$E_{P-\text{TNT}} = \frac{M_{\text{TNT}}}{M} = \left(\frac{Z}{Z_{\text{TNT}}}\right)^3_{Pcst} \tag{8.12}$$

where Z is the scaled distance.

8.2.2.2 Impulse-Based Concept

A similar approach is used to obtain the equivalent mass of impulse (Esparza 1986). The TNT equivalency for impulse is the mass (M) of the explosive that produces a blast with the same incident positive impulse as an equivalent mass of TNT (M_{TNT}) at the same radial distance from each load:

$$E_{I-\text{TNT}} = \frac{M_{\text{TNT}}}{M} = \left(\frac{Z}{Z_{\text{TNT}}}\right)^3_{I.M^{-1/3}cst} \tag{8.13}$$

However, the impulses are reported to be the cube root of the mass, and the equivalent impulse can be obtained by sliding the curves along the diagonal.

This approach to pressure concept is used by Formby and Wharton (1996) and Wharton et al. (2000) to calculate the TNT equivalent for PE4, Super Dopex, nitroguanidine, Powergel 700, Driftex, and Penobel 2. A linear evolution of the TNT equivalent is obtained as a function of the scaled distance.

8.2.2.3 Arrival Time-Based Concept

A similar approach is adopted to obtain the equivalent mass of arrival time. Thus, the TNT equivalency for arrival time is defined as the mass of the explosive that produces a blast with the same arrival time as an equivalent mass of TNT (M_{TNT}) at the same radial distance from each load:

$$E_{Ta-\mathrm{TNT}} = \frac{M_{\mathrm{TNT}}}{M} = \left(\frac{Z}{Z_{\mathrm{TNT}}}\right)^{3}_{Ta.M^{-1/3}cst} \tag{8.14}$$

The curves of the scaled arrival time of explosives can be shifted along the diagonal to determine the arrival time equivalence factor.

Ohashi et al. (2002) and Kleine et al. (2003) described a procedure for calculating the TNT equivalent. This approach is based on knowledge of the shock radius— the time of arrival diagram of the shock wave for the explosive under consideration. These data are then used to calculate the Mach number of the shock and the peak overpressure as a function of distance (Dewey 2005).

8.2.2.4 Chapman–Jouguet State-Based Concept

Cooper (1996) defined the energy equivalent of TNT from the hydrodynamics of the CJ detonation state:

$$E_{\mathrm{CJ-TNT}} = \left(\frac{P_{\mathrm{CJ}}}{2\rho_{\mathrm{CJ}}}\right)\left(\frac{2\rho_{\mathrm{CJ,TNT}}}{P_{\mathrm{CJ,TNT}}}\right) \tag{8.15}$$

where $P_{\mathrm{CJ}} \sim \frac{\rho_0\, D_{\mathrm{CJ}}^2}{4}$, $\rho_{\mathrm{CJ}} \sim \frac{4\rho_0}{3}$ and $\frac{P_{\mathrm{CJ}}}{2\rho_{\mathrm{CJ}}} = \frac{3D_{\mathrm{CJ}}^2}{32}$.

P_{CJ}, ρ_{CJ} and D_{CJ} designate pressure, density, and detonation velocity in the CJ state respectively.

Thus, the equivalent energy of TNT for a given explosive is expressed as:

$$E_{\mathrm{CJ-TNT}} = \frac{D_{\mathrm{CJ}}^2}{D_{\mathrm{CJ,TNT}}^2} \tag{8.16}$$

Jeremie and Bajie (2006) expressed the dependence of pressure P_{CJ} and detonation velocity D_{CJ} on the mass energy E of detonation:

$P_{\mathrm{CJ}} = 2\left(\gamma_{\mathrm{CJ}} - 1\right)\rho_0 E$ and $D_{\mathrm{CJ}} = \sqrt{2\left(\gamma_{\mathrm{CJ}}^2 - 1\right)E}$ where γ_{CJ} is the adiabatic constant of the detonation products and $\rho_0$is the density of the explosive.

8.2.2.5 Explosion Yield-Based Concept

Lannoy (1984) conducted an analysis of 150 incidents that resulted in accidents and fires in the gas, oil, and chemical industries. The results are representative of 23 accidents for which the data are sufficient to yield a calculation of the explosion.

Table 8.4 Severity of explosions—Lannoy (1984)

Range performance (%)	Representative value (%)	Empirical equivalence kg TNT per kg fuel	Corresponding frequency	Cumulative frequency
$0 \leq E_{\mathrm{TNT}} \leq 6$	4	2	0.80	0.80
$6 \leq E_{\mathrm{TNT}} \leq 12$	10	5	0.17	0.97
$12 \leq E_{\mathrm{TNT}} \leq 18$	16	8	0.03	1.00

The database relies on the same pressure-based approach. The TNT equivalent of an explosive or explosive gas mixture is the mass of TNT that causes an explosion with the same pressure field as 1 kg of the explosive. The energy equivalence is defined by the following ratio:

$$E_{\mathrm{TNT}} = \frac{\text{Energy of combustion of fuel}}{\text{Detonation energy of TNT}} \tag{8.17}$$

Lannoy (1984) considered an average value of the combustion energy of $|46.9|$ MJ kg^{-1} for classical fuel/air and $|4.69|$ MJ kg^{-1} for TNT. Thus, a theoretical energy equivalence could be defined as 10 kg of TNT for 1 kg of hydrocarbon. However, the validity of this global value should be determined. For that, Lannoy (1984) suggested an explosion yield energy to establish a comparison with the analysis of the 23 accidents. The explosion yield is defined by the ratio:

$$E_{\mathrm{TNT}} = \frac{\text{Mass of the TNT equivalent} \times 4.69\,\text{MJ kg}^{-1}}{\text{Mass of the released fuel} \times Q\,\text{MJ kg}^{-1}} \tag{8.18}$$

In this equation, Q is the energy released by the complete combustion in air of a unit mass of the product under consideration.

Table 8.4 shows a severity scale for explosions. If an accidental explosion occurs, then the resulting damage can be determined by using a TNT equivalence of 2 for a mass of hydrocarbon within the flammability limits. The resulting damage accounts for 80% of all of the possible damage. The strongest effects are observed for a TNT equivalent of 8 and the probability of observing such a system during actual explosions is greater than 3%.

Thus, the validity of the theoretical energy equivalence can be determined. A value of 10% should be used in a safety analysis to estimate the pressure effects because this value corresponds to a confidence level of 97%. An explosion yield of 10% corresponds to a 5 kg TNT equivalent of 1 kg of hydrocarbon in the atmosphere.

In the case of a high explosive (HE), the identical relation of (8.18) is expressed (U.S. Army Corps of Engineers (2008), for example):

$$M_{e,\mathrm{TNT}} = \frac{H_{\mathrm{d,HE}} \times M_{\mathrm{HE}}}{H_{\mathrm{d,TNT}}} \tag{8.19}$$

where $M_{e,\mathrm{TNT}}$ is the TNT equivalent weight (kg), M_{HE} is the weight of the explosive considered (kg), $H_{\mathrm{d,HE}}$ is the detonation energy of the high explosive considered ($\mathrm{MJ\,kg^{-1}}$) and $H_{\mathrm{d,TNT}}$ is the detonation energy of TNT ($\mathrm{MJ\,kg^{-1}}$).

8.2.2.6 Multi-Parameter Concept

All of the methods used to determine the TNT equivalent are based on the fact that a potential explosion of a large cloud of gas is proportional to the total amount of fuel present in the cloud, whether or not it is within flammable limits. The power of the explosion of a gas cloud is expressed as a charge of equivalent TNT energy. The factor of proportionality is determined from the damage observed in a large number of incidents of exploding gas clouds.

By accounting for the equation for the detonation products, the effects of detonation are influenced by basic parameters such as detonation celerity D, pressure P, detonation energy E, and the number of moles n of the detonating gas. These values can be obtained from thermochemical calculations, and an average value can be obtained. Thus, the following expression is available for fuel and high explosives:

$$E_{\mathrm{TNT}} = k_1 \frac{n_{\mathrm{HE}}}{n_{\mathrm{TNT}}} + k_2 \frac{E_{\mathrm{HE}}}{E_{\mathrm{TNT}}} + k_3 \frac{P_{\mathrm{HE}}}{P_{\mathrm{TNT}}} + k_4 \frac{D_{\mathrm{HE}}}{D_{\mathrm{TNT}}} \tag{8.20}$$

where k_1, k_2, k_3 and k_4 are empirical coefficients that were obtained experimentally (Jeremie and Bajie 2006).

8.3 Applications

Equivalency of TNT is calculated for properties of blast waves generated by the detonation of gaseous mixtures and explosive charges by applying (8.12), (8.13) and (8.14).The evolution of equivalence factors is represented on linear charts. This choice allows more precise values that are applicable for engineers to be obtained.

8.3.1 Application to Gaseous Mixtures

Only the results of Fraunhoffer give an increasing pressure equivalence factor continuously with the scaled distance for spherical charge (Fig. 8.1). Other scale experiments give increasing and decreasing curves, and the larger the scale of experiments with ethylene–air mixtures, the more the evolution of the equivalent is flattened. The pressure equivalence factor for hemispherical charge is less than 1 (Fig. 8.2). However, higher pressure equivalence factors are obtained with spherical charges and small-scale experiments are close to 1 for each gaseous mixture. On

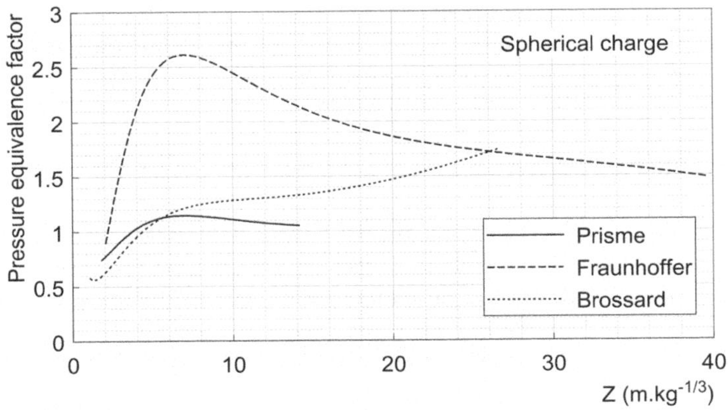

Fig. 8.1 Pressure equivalence factor versus scaled distance in mass for spherical gaseous charge

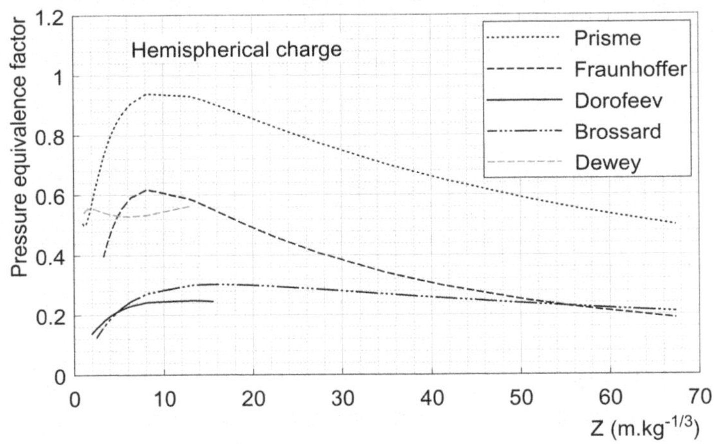

Fig. 8.2 Pressure equivalence factor versus scaled distance in mass for hemispherical gaseous charge

a large scale for hemispherical charge (Dewey), the pressure equivalence factor could be considered to be constant (0.54) with regard to the scaled distance of the study.

Considering the hemispherical charge (Fig. 8.3), large- scale experiments cannot achieve an arrival time factor of 1. The arrival time equivalence factor is higher than the unit in the case of Fraunhoffer's results and corroborates the previous results (Fig. 8.3) for hemispherical charges. A long-scale distance is necessary to ensure the convergence of the arrival time factor with hemispherical charges on a large scale (Dewey 2005). The arrival time equivalence factor converges toward an asymptote at 1 in the intermediate field with spherical charges (Fig. 8.4), whatever the size of the experiments.

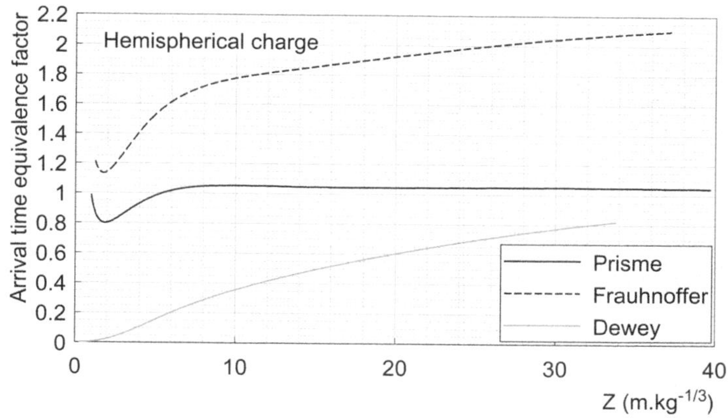

Fig. 8.3 Arrival time equivalence factor versus scaled distance in mass for hemispherical gaseous charge

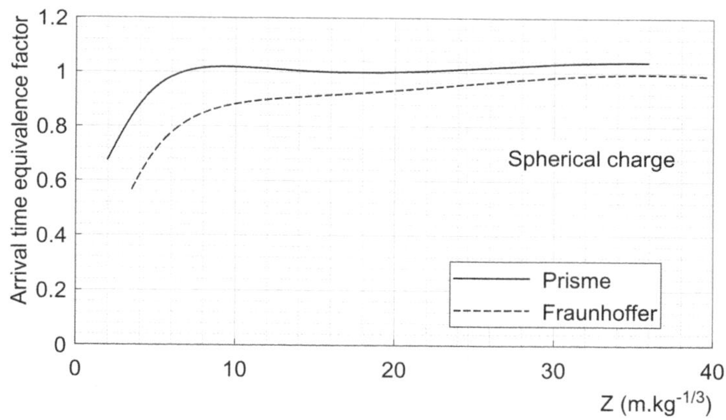

Fig. 8.4 Arrival time equivalence factor versus scaled distance in mass for spherical gaseous charge

The impulse equivalence factors are very low, particularly for larger volumes. Hence, for a given scaled impulse the equivalent mass of TNT is smaller the greater mass of gas (Fig. 8.5).

8.3.2 Applications to High Explosives

The equivalence factor is calculated for the explosives studied in Chap. 7. The results for pressure are reported in Fig. 8.6.

The spherical charge of composition B presents a TNT equivalency on the pressure approximately equal to 1 on the domain 2.4–6 m kg$^{-1/3}$. Beyond this

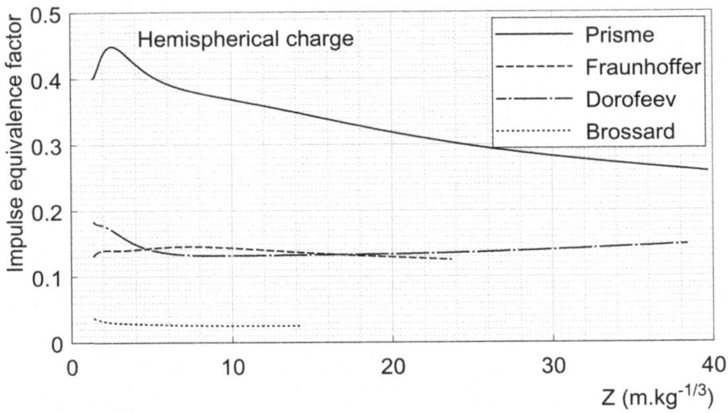

Fig. 8.5 Impulse equivalence factor versus scaled distance in mass for hemispherical gaseous charge

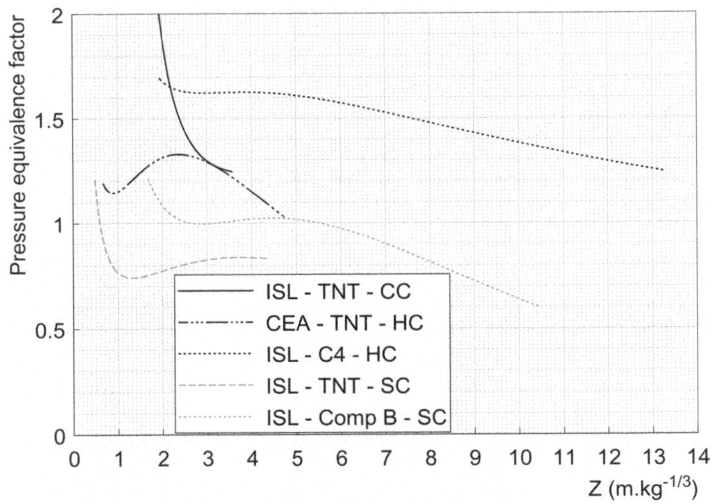

Fig. 8.6 Pressure equivalence factor versus scaled distance in mass for high explosives

value, the TNT equivalency decreases with scaled distance. The constant range is consistent with the equivalence factor of 1.11 available in the literature (Swisdak 1975; U.S. Department of Energy 1980; Hokanson et al. 1982; IATG 2015) for a pressure 0.35–3.5 bar. This pressure domain corresponds to a scaled distance 1.8–5 $\mathrm{m\,kg^{-1/3}}$ for the reference curve of TNT overpressure (U.S. Army Corps of Engineers 2008). Remennikov (2007) reported a TNT equivalent of 1.148 for a classical composition B (60% RDX and 40% TNT).

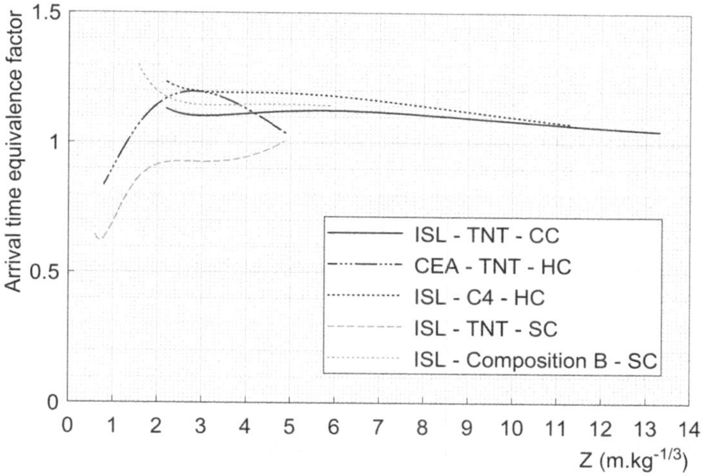

Fig. 8.7 Arrival time equivalence factor versus scaled distance in mass for high explosives

The pressure equivalence factor decreases monotically with scaled distance Z for hemispherical charge C4 and is higher than the values reported by Krauthammer (2008). He gives an equivalent mass for pressure of 1.20 for pressure 0.7–13.8 bar ($Z = 3 - 0.9\,\mathrm{m\,kg^{-1/3}}$) and 1.37 for pressure 13.8–20.7 bar (or $Z = 0.9 - 0.6\,\mathrm{m\,kg^{-1/3}}$). The equivalency value of 1.37 corresponds to the published factor by Parmentier (1993) and Swisdak (1975), IATG (2015). Bogosian et al. (2016) have recently recommended laws of pressure and impulse equivalence factor versus scaled distance with two-sigma confidence bounds.

The spherical and small charge of TNT (ISL) leads to a decreasing pressure equivalence factor in the near field and globally constant and lower than the unit in the intermediate field. In contrast to the more important and hemispherical charge used by the CEA, which gives an equivalence factor greater than 1.

For the cylindrical charge of TNT, the pressure equivalence factor decreases greatly and is reported here at the limited scaled distance of the investigation.

Considering the arrival time equivalence factor in the function of scaled distance (Fig. 8.7) a flat evolution around 1–1.2 is obtained for the cylindrical charge of TNT, the hemispherical charge of C4, and the spherical charge of composition B. Hence, the sensitivity of the explosive is less important for the arrival time equivalence factor than for the pressure equivalence factor.

8.4 Limitations of the TNT Equivalent Method

Regarding the various methods, Gelfand (2004) proposed defining a mean value. This mean value defines the TNT equivalent energy by using the ratio of the detonation energy of an explosive to the detonation energy of TNT. However, the determination of the exact value of the detonation energy is difficult, and its value ranges from −3.21 to −4.918 MJ kg^{-1} (ratio of 1.5) based on the rule of decomposition and the thermodynamic data that were selected. This implies that the TNT equivalent values are affected by the choice of good data. Besides, as explained by several authors (Kleine et al. 2003; Dewey 2005; Bogosian et al. 2016; Rigby and Sielicki 2014 for example), the TNT equivalence depends on the properties of the blast wave and the scaled distance parameter. Considering the different possible geometries of high explosives (spherical, cylindrical, orthorhombic), the TNT equivalency varies greatly with the direction and the distance. As Cooper mentioned (Cooper 1994) the experimental determination of TNT equivalency is subject to errors.

A disparity of TNT equivalency values is obtained as a function of the approach considered for the calculation.

In practice, the calculation of the TNT equivalent can evaluate the effects of an explosion. However, the abacus was constructed for ideal blast waves, point sources, and for explosions that occur in free air. If the explosion takes place in a complex environment like a closed zone, urban area or industrial facilities, it becomes impossible to define a TNT equivalency.

The TNT equivalent method is based on the approximation of a point source for an explosion. Despite the energy of a gaseous explosion being greater than the energy of HE detonation, the equivalence factors are lower. In the case of accidental gaseous explosion, the energy released is not limited to a point source and the cloud is not homogeneous or hemispherical or spherical. The leakage of gas is usually caused by a rupture of the tank, which has an effect on the distributed energy. Moreover, the isopressures are considered to be spherically symmetric from the point source and are assumed to be at the origin of the blast wave; therefore, there is no relationship to the geometry of the cloud. To the knowledge of the authors, only Lannoy (1984) proposed an approach to account for the nonspherical shape of the curves of isodamage. Lannoy considered the minimum and maximum distance for which the same damage was observed that was associated with the minimum and maximum explosive yields. The nonspherical shape of the isodamage curve may be due to the geometry of the site where the explosion occurred, the effect of the wind in the case of a gas release, and the formation of a gas cloud before initiation.

Despite these drawbacks, the TNT equivalent method for the determination of the blast resulting from the detonation of an explosive is the simplest, the fastest, and cheapest method of obtaining a first analysis of blast properties.

If the TNT equivalence is needed to determine damage and injury criteria, then it is probably best to use an impulse comparison. Unfortunately, this requires a knowledge of the pressure–time histories of the blast wave from an explosive, over a wide range of distances. This is much more difficult to obtain than the peak pressures, which can be determined from a knowledge of the shock time-of-arrival only, and it is for this reason that TNT equivalence is usually based on peak pressure rather than on impulse.

The difficulty of estimating a single value of TNT equivalence is confirmed by Peugeot et al. (2006). He describes the influencing parameters on the equivalent: the energetic composition of the explosive, the distance, and the geometry of the charge are described as influencing parameters on the TNT equivalent. If no measurements are made or analyses carried out, then in agreement with Allied Ammunition Storage and Transport Publication (2003), an energetic equivalent factor of TNT of 1.4 can be suggested and leads to a conservative estimation of TNT mass for an unknown explosive.

References

Akhavan, J. (2004). *The chemistry of explosive* (2nd ed.). Cambridge: Royal Society of Chemistry.

Allied Ammunition Storage and Transport Publication (AASTP)-4. (2003). Explosives Safety Risks Analysis, Part II - AC/258(ST)WP/221.

Baker, W. E. (1973). *Explosions in air*. Austin: University of Texas Press.

Bogosian, D., Yokota, M., & Rigby, S. (2016, September). TNT equivalence of C-4 and PE-4: a review of traditional sources and recent data. In *MABS 24, 24th International Symposium on the Military Application of Blast and Shock, Proceedings, Halifax, Canada*.

Cooper, P. W. (1994, July 24–29). Comments on TNT equivalence. In *20th International Pyrotechnics Seminar Colorado Springs, Colorado*

Cooper, P. W. (1996). *Explosives engineering*. Weinheim: Wiley-VCH.

Dewey, J. M. (2005). The TNT equivalence of an optimum propane-oxygen mixture. *Journal of Physics D: Applied Physics, 38*, 4245–4251

Esparza, E. D. (1986). Blast measurements and equivalency for spherical charges at small scaled distances. *International Journal of Impact Engineering, 4*(1), 23–40.

Filler, W. S. (1956). Post-detonation and thermal studies of solid high explosives in a closed chamber. *Combustion of Explosives and Solid Propellants, 6*, 648–657.

Formby, S. A., & Wharton, R. K. (1996). Blast characteristics and TNT equivalence values for some commercial explosives detonated at ground level. *Journal of Hazardous Materials, 50*, 183–198.

Gelfand, B. (2004, April). Translation from Russian to English The Book "Blast effects caused by explosions" authorised by B. Gelfand and M. Silnikov. United States Army, European Research Office of the U.S. Army, London, England, Contract Number N62558-04-M-0004.

Hokanson, J. C., Esparza, E. D., Baker, W. E., Sandoval, N. R., & Anderson, C. E. (1982). Determination of blast loads in the DWF, I, II, SwRI - 6578. San Antonio.

Jeremie, R., & Bajie, Z. (2006). An approach to determining the TNT equivalent of high explosives. *Scientific-Technical Review, 56*(1), 58–62.

Kinney, G. F., & Graham, K. J. (1985). *Explosives shocks in air* (2nd ed.). Berlin: Springer.

Kleine, H., Dewey, J. M., Oashi, K., Mizuka, T., & Takayama, K. (2003). Studies of the TNT equivalence of silver azide charges. *Shock Waves, 13*, 123–138.

Krauthammer, T. (2008). *Modern protective structures*. Boca Raton: CRC Press, Taylor & Francis Group.

Lannoy, A. (1984). Analyse des explosions air-hydrocarbure en milieu libre: Etudes déterministes et probabiliste du scénario d'accident. Prévision des effets de surpression, Bulletin Direct. Etudes et Recherche EDF. A4 (Published in French).

Meyer, R., Köhler, J., & Homburg, A. (2007). *Explosives* (6th ed.). Weinheim: Wiley-VCH Verlag GmbH.

Ohashi, K., Kleine, H., & Takayama, K. (2002). Characteristics of blast waves generated by milligram charges. In F. Lu (Ed.), *23rd International Symposium on Shock waves, Fort Worth, USA* (pp. 187–193).

Omang, M., Christensen, S. O., Borve, S., & Trulsen, J. (2009). Height of burst explosions: a comparative study of numerical and experimental results. *Shock Waves, 19*(2), 135–143.

Parmentier, G. (1993, Juin 18). Synthèse des résultats expérimentaux relatifs aux détonations d'explosifs sphériques. Institut Franco-Allemand de Recherche de Saint-Louis - R 113/93 - Contrat 91.02.178/ ETBS-CETAM (in French).

Peugeot, F., Deschalbault, E., & Péron, P. F. (2006, October). *L-132 TNT equivalency: misconceptions and reality*. Brussels: Munitions Safety Information Analysis Center, MSIAC Unclassified.

Pförtner, H. (1977). Gas cloud explosions and resulting blast effects. *Nuclear Engineering and Design, 41*, 59–67.

Remennikov, A. M. (2007). *The state of the art of explosive loads characterisation*. Wollongong: University of Wollongong. http://www.aees.org.au

Rigby, S. E., & Sielicki, P. W. (2014). An investigation of TNT equivalence of hemispherical PE4 charges. *Engineering Transactions, 62*(4), 423–435.

Scilly, N. F. (1995). Measurement of the explosive performance of high explosives. *Journal of Loss Prevention in the Process Industries, 8*(5), 265–273.

Swisdak, M. M. J. (1975). Explosion effects and properties- Part I - Explosion effects in air, report NSWC/WOL/TR-116.

Tongchang, Y., Menchao, Y., & Jianling, W. (1995). Determination of heats of detonation and influence of components of composite explosives on heats of detonation of high explosives. *Journal of Thermal Analysis, 44*, 1347–1356.

UN SaferGuard User Survey. (2015). International Ammunition Technical Guideline IATG 0.180, Formulae for ammunition management (2nd ed.) 2015-02-01.

U.S. Army Corps of Engineers, Naval Facilities Engineering Command, Air Force Civil Engineer Support Agency. (2008). Technical Manuals, Unified Facilities Criteria (UFC), Structures to resist the effects of accidental explosions, UFC 3-340-02.

U.S. Department of Energy. (1980). A Manual for Prediction of Blast and Fragment Loadings on Structures. DOE/TIC - 11268.

Wharton, R. K., Formby, S. A., & Merrifield, R. (2000). Airblast TNT equivalence for a range of commercial blasting explosives. *Journal of Hazardous Materials, A79*, 31–39.

Chapter 9
Small-Scale Blast Wave Experiments by Means of an Exploding Wire

Oren Sadot, Omri Ram, Eliram Nof, Eytan Kochavi, and Gabi Ben-Dor

9.1 Introduction

Understanding how blast waves interact with structures has become increasingly important over the past decades, especially in light of the increased threats of terror attacks. Blast–structure interactions depend on two processes: the dynamic loads imposed by the explosion and the structural response to these loads. The time scales of these two processes differ markedly, with that of blast propagation in the structure being significantly shorter than that of the response of the structure to the blast due to its large mass. In most cases, the difference between these two time scales justifies neglecting the effect of the motion of the structure on the propagation of the blast wave in the structure. Details concerning the structural response to the load can be found in Smith and Hetherington (1994).

In this chapter, we introduce a unique experimental tool that enables investigating the blast wave imposed load on a structure using different types of diagnostics systems.

Some studies on the dynamic loads associated with explosions are full-scale experiments, but such resource-intensive tests usually have inherent, obvious difficulties (Needham 2010; Dewey et al. 1977). Although numerous measuring techniques, including those based on high-speed photography, have been developed over the past few decades to measure the effects of blast waves on structures, their applicability is limited due to the sheer destructiveness of large-scale explosions. In fact, it is virtually impossible to monitor blast wave propagation and dynamic gas effects in full-scale urban scenario experiments using high-speed photography (Needham 2010; Dewey 2001). Instead, the data gleaned from full-scale urban

O. Sadot • O. Ram • E. Nof • E. Kochavi • G. Ben-Dor (✉)
Protective Technologies R&D Center, Faculty of Engineering Sciences, Ben-Gurion University of the Negev, Beer Sheva, Israel
e-mail: bendorg@bgu.ac.il

I. Sochet (ed.), *Blast Effects*, Shock Wave and High Pressure Phenomena,
https://doi.org/10.1007/978-3-319-70831-7_9

scenarios are limited to measurements from a small number of pressure transducers positioned in key locations inside the structures. But such data can help only in partially understanding blast wave flow patterns and propagation of blast waves inside the structures.

A widely used alternative technique to investigate the effects of an explosive event on structures exploits small-scale tests under laboratory conditions. Under controlled conditions, small-scale explosions can be easily created using chemical explosives such as silver-azide (Kleine et al. 2005; Hargather and Settles 2007). Such experiments utilize small charges that typically range from a few milligrams for a laboratory test (Kleine and Takayama 2004; Kleine et al. 2005; Hargather and Settles 2007; Cheval et al. 2010) to a few kilograms for a scaled-down field test (Neuberger et al. 2007). A variety of diagnostic methods, including pressure transducers and strain gauges placed in the vicinity of the explosion and on the tested structures (Reichenbach and Neuwald 2000) and optical diagnostic systems, can be used to measure the dynamic flow field and the structure response. Pressure transducers and strain gauges record the time history of the load developed by the explosion and the structural response to the load, while optical systems, such as schlieren or shadowgraph techniques, are used to monitor the blast wave propagation (Kleine and Takayama 2004; Hargather and Settles 2007; Reichenbach and Neuwald 2000; Settles 2001). The advantages of such small-scale experiments include their affordability (Kleine and Takayama 2004; Kleine et al. 2005) and the ease with which a large number of experiments can be safely conducted to verify and validate results obtained from numerical simulations. The validated results, in turn, can provide a better understanding of how wave propagation, flow pattern, and loads behave in actual explosions. Indeed, it has been shown that despite the obvious differences between the properties of small vs. large explosive charges, relatively good predictions can be made based on the information acquired in small-scale experiments combined with various scaling laws (Needham 2010; Dewey 2001; Hargather and Settles 2007).

One of the most comprehensive collections of experiments to study internal and external explosions in small-scale scenarios was published by Reichenbach and Neuwald (2000). In their examples, gram-scale explosive charges were used to study blast effects on buildings, stairways, tunnels, and other structures found in urban scenarios. Pressure transducers were used to measure the pressures produced by the explosions and shadowgraph or schlieren photography was used to monitor blast wave propagation in and around the structures. Likewise, the experiments of Smith et al. (1992) with small-scale internal and external explosions showed that the results can be used to obtain blast loading data in complex structures, such as tunnels and rooms. For example, they showed that a relatively good assessment of peak overpressure can be obtained using the well-known Crantz-Hopkinson scaling laws.

Small-scale blast experiments have been used to calibrate and validate numerical hydro-codes used to study a wide range of problems related to blast waves. Jiang et al. (1998) studied the micro-blast wave generated by a near instantaneous release of energy from a pulsed Nd:glass laser focused on a point several tens of microns in diameter. Blast wave visualization by double-exposure holographic interferometry and numerical simulation showed that a spherical micro-blast wave is a valid tool for simulating the blast waves created in large-scale explosions. Rose and Smith (2002) studied the dependence of the positive and negative phase impulses of blast waves from explosive detonations on the geometric features of city streets (i.e., tall buildings on both sides of a street). Their small-scale experiments (1/40th scale), designed to model a 1000-kg TNT spherical explosive charge exploded at street level, were performed with 11.09 g DEMEX 100 type explosive, and their results were used to validate numerical stimulations (the three-dimensional blast simulation code "Air3d" was used). Zyskowski et al. (2004) used small-scale experiments together with numerical simulations to assess the effects of an accidental explosion of a confined hydrogen-air gas mixture cloud on nearby structures. They created a confined explosion by electrically igniting a hemispherical soap bubble filled with a mixture of hydrogen and oxygen and situated in a closed box. Reflected pressure histories on the box walls, measured at different distances from the explosion, were used to validate a numerical solution obtained using the AUTODYN commercial solver. The authors found better correlations for normal than for non-normal reflection areas and that the pressure measurements from the simulations were lower than those in the experiments.

Establishing our experimental design on the essence of the small-scale experiment, we developed an experimental tool to investigate the dynamic loads that are imposed on a structure by internal and external blast waves. We created small-scale explosions using the exploding wire approach, in which an applied high current pulse causes the abrupt evaporation of a thin metal conductor. Popular since the 1920s, this approach has been used mostly to study the electrical and physical phenomena associated with the behavior of the wire (Chase and Moore 1959). Nowadays, exploding wires are used to study the effects of high currents on, among other phenomena, conductors and the plasma state of matter. The strong blast wave that accompanies the exploding wire phenomenon has also been investigated by Bennet (1959), Buntzen (1962), and Higashino et al. (1991). Those studies used capacitor banks and fast, high current switches to evaporate the thin conductive wire and create the blast wave in an earth, a gas (air), or a liquid (water) medium. To monitor the blast propagation and pressure history, a high-speed, schlieren-based photography system and pressure transducers were used. To compare the experimental results to the numerical ones, the exploding wires were modeled in the simulations as high density, high temperature columns of air modeled as an ideal gas. Good agreement was found between the numerical calculations and the flow fields produced by the exploding wires (Higashino et al. 1991).

Based on the above studies and being aware of the potential of the exploding wire system for blast wave research, we built a similar system and developed the corresponding research methodology. The newly built system and its complementary numerical model enabled us to deepen our understanding of the complex blast–structure interaction phenomenon, predict blast flow behavior, and evaluate quantitatively the reliability of the numerical code.

The present chapter is organized as follows: the experimental system and the calibration process are first described, after which we demonstrate the effectiveness of the research methodology, present two test cases demonstrating the capabilities of the exploding wire facility to investigate the interaction of small-scale cylindrical and spherical blast waves with structures.

9.2 Experimental Setup

9.2.1 Exploding Wires

When a short, high-current pulse passes through a thin conducting wire, the wire undergoes a rapid Joule heating and evaporates. The current is created by discharging high voltage from a capacitor through the thin wire. Two basic circuits, the charging and discharging circuits, are connected to the capacitor, which is charged by a high-voltage power supply. The discharge circuit is connected to the capacitor through a fast-opening high-voltage switch whose closure connects the capacitor to the wire. This action permits the stored charge to flow through the thin wire causing a very high current pulse. The wire then undergoes a very fast Joule heating, which causes it to liquefy and vaporize with virtually no change in its volume. This leads to the creation of a very hot and dense vaporized metal column that starts to expand rapidly. The rapid expansion of the hot vaporized metal column drives a strong blast wave. A detailed explanation of this phenomenon can be found in Reithel et al. (1962).

The amount of charge stored in the capacitor is determined by the charging voltage, which is set before the experiment. If the energy is too low the wire exhibits "fuse"-like behavior that will not produce a blast wave. On the other hand, too much energy will produce non-repeatable results, and some electric energy will remain in the capacitor. Under correct working conditions, the exploding wires produce repeatable blast waves. The two energy limits must thus be found for each wire configuration. Exploding wires, by their nature, produce cylindrical blast waves, but as will be shown subsequently, we successfully extended the use of the exploding wire phenomenon to generate spherical explosions that drive spherical blast waves, which more closely resemble the spherical explosions generated by high explosive charges. In our system, spherical explosions were obtained by tying a knot in the wire (whose ends were held between the two electrodes (Fig. 9.1)) that was then exploded as discussed above. The wire was isolated to prevent an electrical short circuit. Thus, the small volume explosion produced by our system formed a perfectly spherical blast wave.

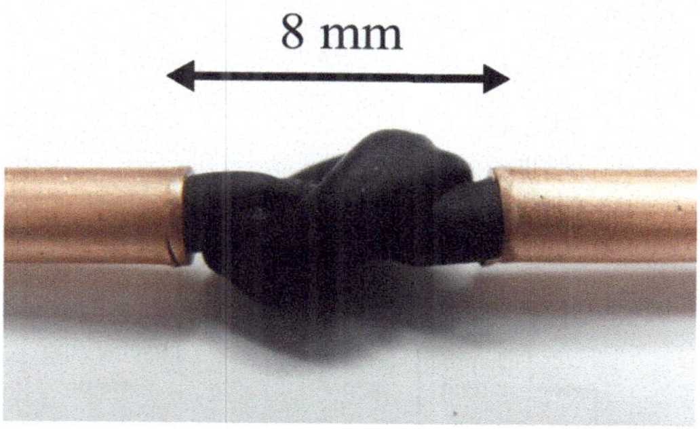

Fig. 9.1 Knot configuration—isolated knot placed between the two electrodes

9.2.2 Experimental Apparatus

The basic components of the experimental apparatus are presented in Fig. 9.2. The main system was contained in a metal frame measuring 0.8 m × 0.8 m × 2.2 m. The box was divided by a partition, such that the lower part contained the electrical system (capacitor, spark gap, and safety circuit) and the upper part housed the hydrodynamic experiment. A remotely controlled vacuum pump was used to initiate the high-current switch by pumping out the isolating gas that initially filled the box. Safety measures in the laboratory included entryway controls and interlocks.

9.2.3 Electrical Outline

The electrical system (shown in Fig. 9.2) contained three circuits, one for charging and two for discharging (one through the wire and the other through the discharge resistor). The charging circuit included a 200-μF capacitor, a charging resistor, and a 20-kV, 20-mA high-voltage power supply. The charging process was controlled remotely, and the capacitor voltage was continuously monitored. The first discharging circuit included the capacitor, the fast switch (spark gap), and the wire to be exploded. The spark gap was designed to permit high current flow in a vacuum but to be an isolator when filled with SF_6 at atmospheric pressure. By pumping the SF_6 out of the spark gap, the current stored in the capacitor could flow through the discharge circuit and explode the wire. The resistance and inductance of the discharging circuit were $R = 14$ mΩ and $L = 0.24$ μH, respectively. This created with the capacitor an under-damped RLC circuit.

To enable side and top views of the experimental apparatus, two optical high-speed flow visualization configurations were used. For the side view, two transparent PMMA windows were mounted on the experimental frame (see Fig. 9.3a). A parallel beam of light aimed through the experimental region created a shadowgraph projection of the shock fronts on a semi-transparent screen. For the top view, a beam of light entering the test chamber from the top passed through the experimental

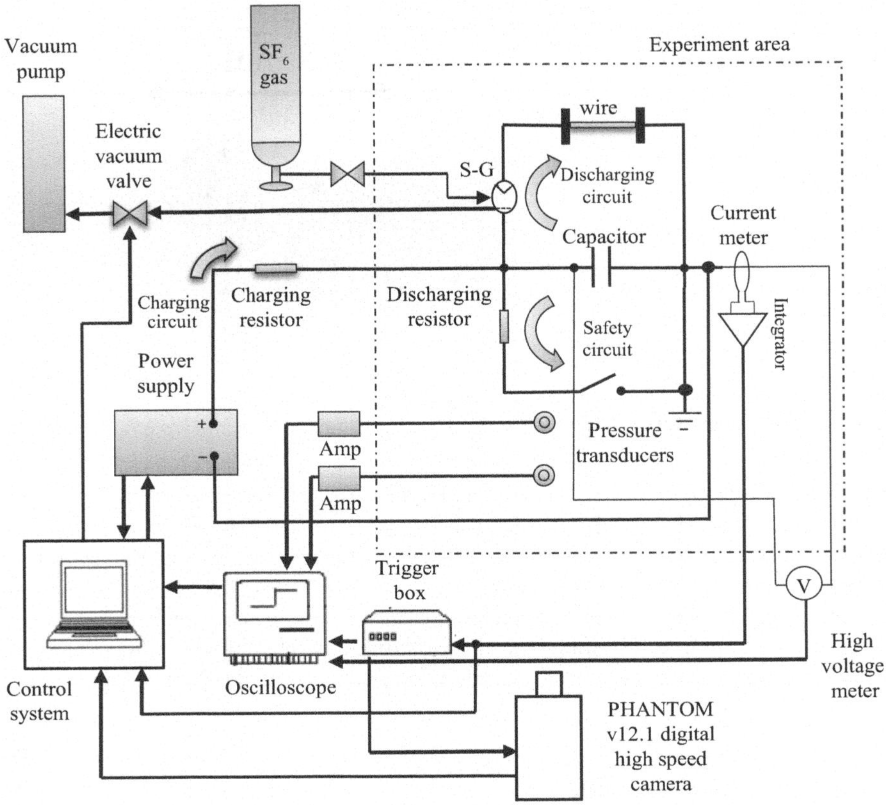

Fig. 9.2 Schematic illustration of the components of the exploding wire system in the BGU shock wave laboratory

region and created a shadowgraph projection on the floor of the test chamber (see Fig. 9.3b). The projected images were captured by the high-speed camera placed beside the test chamber via a planar mirror mounted above the explosion area. The light source for the shadowgraph setups was a double-frequency pulsed Nd^{+3}:YAG laser (532 nm) that could produce 240-ns long pulses with a repetition rate of up to 50 kHz and pulse energy of approximately 2.4 mJ. The light from the remotely situated laser entered the system room via a multimode optical fiber. The expanding light emitted from the fiber was transformed into a parallel beam with a 250-mm parabolic mirror and projected through the test area onto a screen. A high speed digital camera (Phantom v12.1) capable of capturing images at a rate of 20,000 fps with a resolution of 512 × 512 pixels was used to monitor the entire event. A 532-nm band-pass filter (10-nm wide) mounted in front of the camera was used to block the self-emission light from the wire explosion. The laser pulse rate was measured with a photodiode placed near the laser output.

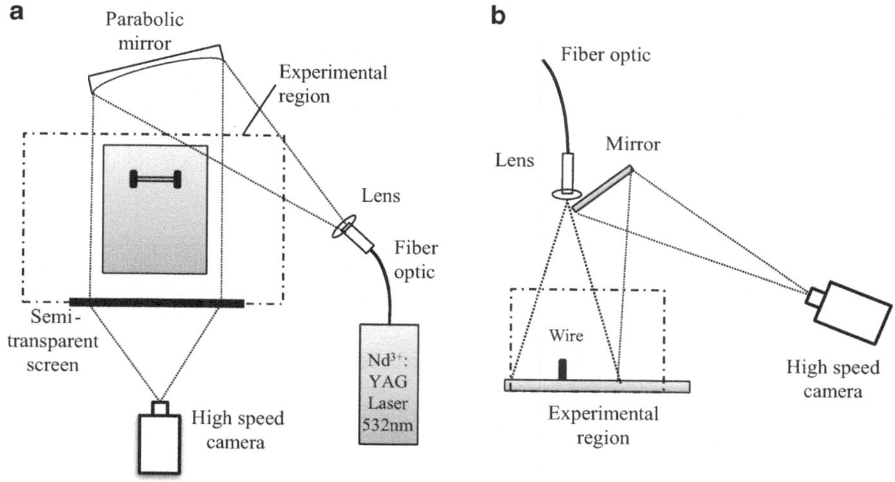

Fig. 9.3 The high-speed shadowgraph setups for (**a**) side view and (**b**) top view

The pressure history recording system was based on Kistler 211B2 and 211B3 piezoelectric transducers, which were calibrated in-house using a shock tube system. The pressure transducers mounted inside the test chamber were connected to the acquisition system, which consisted of a Kistler signal conditioner model 5118B2 and a digital oscilloscope. The current passing through the exploding wire was measured with a specially built and calibrated Rogowski coil placed around one of the conductors between the capacitor and the exploding wire. The measurement system, based on two Lecroy 314A wave-jet 100 MHz, 1 GS/s oscilloscopes, was capable of acquiring data from 8 channels simultaneously. An external trigger box was built to trigger both the high-speed camera and the acquisition system. The latter was triggered by a voltage drop in the capacitor or by the current pulse measured by the Rogowski coil.

9.2.4 Image Processing

To enhance image quality, a background image that was acquired before the shock wave was generated was subtracted from each consecutive image. This procedure enabled us to eliminate nonuniform backlight distribution and some imperfections in the windows to produce a clean image in which the wave front, flow pattern, and fragments are more pronounced. Figure 9.4, an example of the results of this procedure, shows the effect that background subtraction can have on an untreated image.

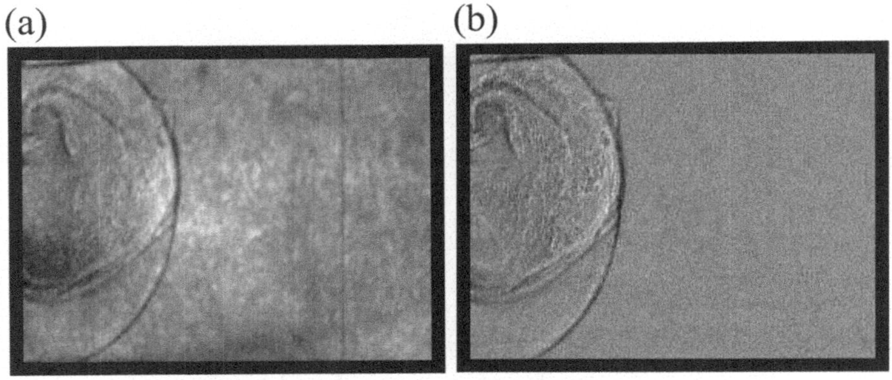

Fig. 9.4 An example of image processing results: (**a**) unprocessed image, (**b**) the same image after processing, i.e., subtracting the background image from (**a**)

9.3 Calibration

The first stage of the experiment was designed to determine the ideal operational conditions of the system. For the exploding wire configuration, two 70-mm long copper wire diameters were used, 0.9 mm and 1.1 mm. For the exploding knot configuration, a 1.25-mm wire diameter was used to ensure that a larger mass of copper was available for the explosion. The voltage and the current signals were monitored to assess the explosion process, and suitable working points, identified through trial and error, were characterized by the complete explosion of the wire with no energy left stored in the capacitor. The charging voltages for the working points were found to be 6 kV and 4.8 kV for the 1.1 mm and 0.9 mm wires, respectively, and for the exploding knot the used voltage was 4.5 kV. To measure the strength of the explosion, a pressure transducer was mounted on a movable stand placed in front of the explosion at distances ranging from 40 to 150 mm. For the experiments with exploding knots, the measurements were compared to CONWEP generated results (CONWEP 1992). In an iterative process it was found that the equivalent amount of TNT that gave similar results was 60 mg (in spherical charge geometry). A numerical simulation had to be used in order to find the energy embedded in the blast wave in the exploding wire experiments. An axis-symmetric numerical simulation was solved using the MSC/Dytran solver (2008r1). To simulate a cylindrical explosion, a high density cylindrical bubble with high internal energy was placed in the explosion position under ambient conditions. We used an iterative process to determine the conditions that promoted generation of the blast wave by the exploded wires. The results obtained from the wire experiments were fitted to the theoretical cylindrical model of Lin (1954) given by Eq. (9.1),

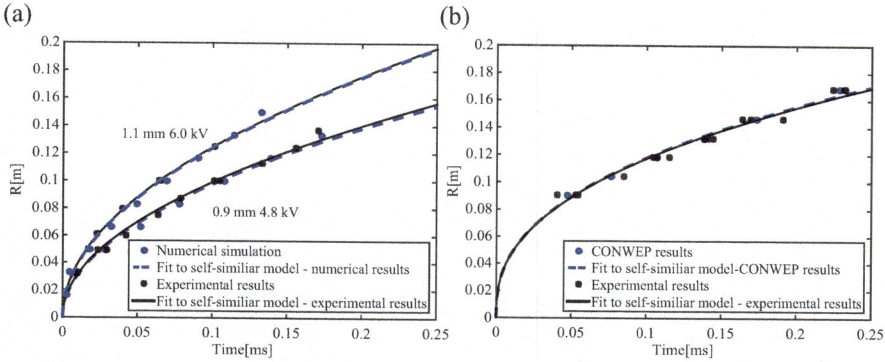

Fig. 9.5 Comparison between the numerical and the experimental results that were used in the calibration procedure

Table 9.1 The charge energies and the energies embedded in the blast waves that were found using the calibration procedure for the three working points

Experimental configura-tion	Charging voltage (kV)	Capacitor charged energy (kJ)	Calculated embedded energy
1.25-mm knot	4.5	2.25	0.3 kJ
1.1-mm wire	6.0	3.6	1.14 kJ/m
0.9-mm wire	4.8	2.3	0.33 kJ/m

while those obtained from the knot experiments were fitted to the theoretical model of Taylor (1950) given in Eq. (9.2):

$$R(t) = A\,(E, \rho, \gamma) \cdot t^{\frac{1}{2}} \tag{9.1}$$

$$R(t) = B\,(E, \rho, \gamma) \cdot t^{\frac{2}{5}} \tag{9.2}$$

where $R(t)$ is the radial position of the incident blast wave, t is the time, and $A(E, \rho, \gamma)$ and $B(E, \rho, \gamma)$ are constants that depend on the energy of the explosion ambient density and the heat capacities ratios. The energies found to fit the wire experiments best were 0.33 and 1.14 kJ for the lower and higher working points, respectively. Comparisons between the experimental and numerical results of the exploding wire and knot experiments are presented in Fig. 9.5a, b. Table 9.1 summarizes the charge energies and the energies embedded in the blast waves that were found using the calibration procedure for the three working points.

Once the energies embedded in the explosions have been found, the symmetry of the spherical blast created by the knot was examined. The blast wave fronts images taken from side and top views were fitted to a circle. The fitting showed that the exploding knot produced a perfectly spherical explosion (see Fig. 9.6).

The head-on pressure history was measured by a pressure transducer flush mounted on a perpendicular stand at a height of 70 mm above the floor, which was the height of the explosion. The distance between the pressure transducer and the explosion was 118.5 mm. The setup configuration is shown in the inset in Fig. 9.7. The measured reflected pressure history from a pressure transducer was compared to

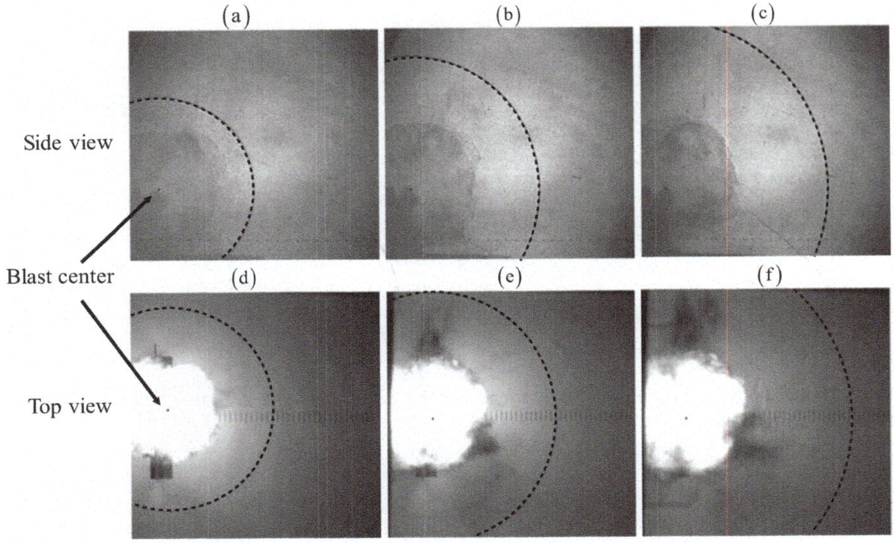

Fig. 9.6 Results of the exploding knot experiment: (**a–c**) side view; (**d–f**) top view. The fitted circles are superimposed on the images and are plotted as dashed lines. The dots shown in each of the figures (**a–f**) represent the blast center as found by the fitted circle

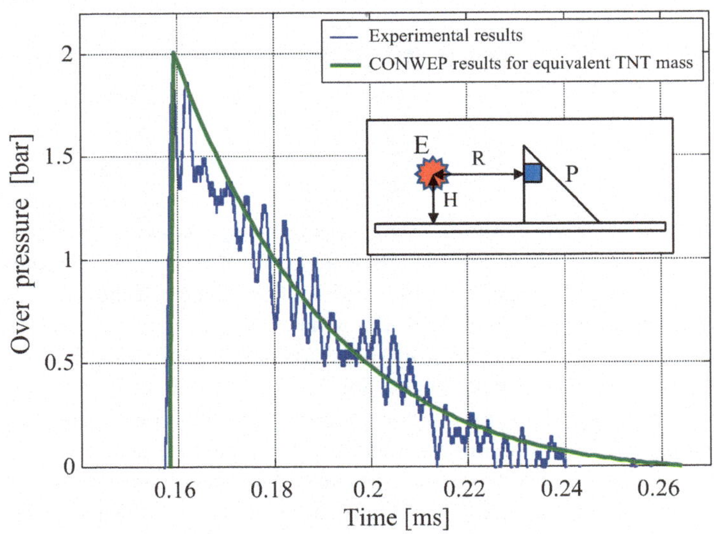

Fig. 9.7 A comparison of pressure histories between the CONWEP and the experimental results. Inset: measurement configuration: P-pressure transducer, E-explosion, $R = 118.5$ mm, $H = 70$ mm

a CONWEP generated pressure profiles. As can be seen in Fig. 9.7 good agreement was found between the pressure histories generated by the exploding knot and the CONWEP results.

9.4 Results

In the following, two experimental studies demonstrating the capabilities of the exploding wire facility to investigate the interaction of small-scale cylindrical and spherical blast waves with complex structures are presented.

9.4.1 Case A: The Interaction of a Cylindrical Blast Wave with a Two-Story Building

To demonstrate the suggested methodology to predict blast loads and to validate the numerical simulation, a simple but nontrivial experiment was designed. It was shown that the experimental apparatus is capable of producing 2D and 3D blast waves. To demonstrate the effectiveness of the experimental system a simple 2D case was tested. The experimental results were compared to the numerical results of a simple 2D numerical solver in which a simple Euler scheme was used to help explain the discrepancies observed between the experimental and numerical results. A comparison of the experimental results to the numerical model solved under the same conditions could enable one to deduct the strengths and the weaknesses of the numerical code.

The following sections present the test model, experiments, and numerical model, and the comparison between the results reveals the effectiveness of using such a system.

9.4.1.1 Numerical Approach

As mentioned, this study aimed at demonstrating the effectiveness of using an exploding-wire-based experimental system to predict blast loads and to validate numerical simulations. To that end, we chose, as a test case, to solve a numerical model using the MSC/Dytran commercial solver (2008r1), which is capable of solving fully coupled fluid and solid conservation laws in both Eulerian and Lagrangian mesh grids. For the purposes of this study, a second order solver was chosen that solved the Euler equations for the fluid phase, and the solid structures were assumed to be infinitely rigid. Furthermore, in the numerical model the exploding wire was replaced by a high density, high energy cylindrical gas bubble situated in the blast location. The method of using a compressed hot gas bubble to generate the exploding wire conditions in simulations was examined by Higashino et al. (1991), who found that it provided very good results. The hot gas bubble and the surrounding atmosphere were modeled as an ideal gas, and a simple γ-law was used for the equation of state of the gas. Special considerations about the numerical grid element size were made to ensure that the blast waves generated in the simulation were correctly modeled. The mesh was refined to the point at

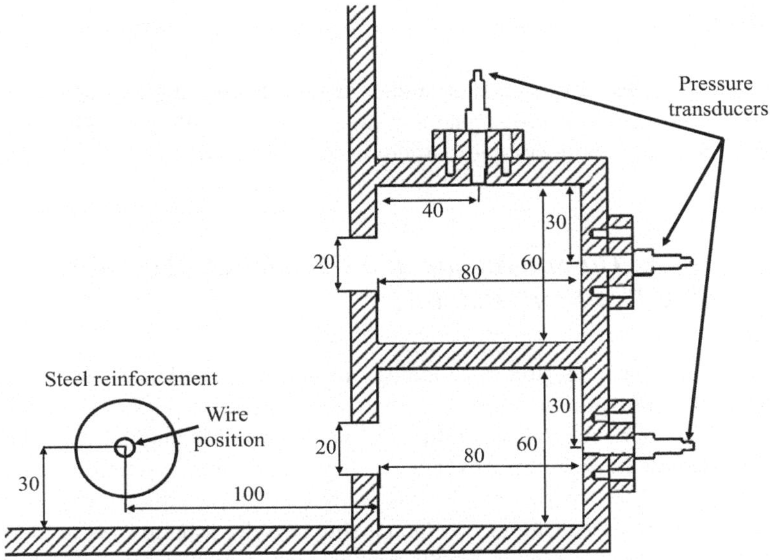

Fig. 9.8 Schematic illustration of the small model (dimensions are in mm)

which further refinement had no influence on the solution. To simplify the numerical solution and maintain the accuracy that is required to ensure good comparisons with the experimental results, the dimensionality of this test case was chosen as two.

9.4.1.2 The Test Model

The exploding wire produced a cylindrical blast wave. Its confinement between two parallel planar walls preserved the two-dimensional nature of the blast wave until later times. The small-scale model presented in Fig. 9.8 was designed as a generic two room structure, in which the rooms are placed in a two-story configuration. The room dimensions in the model were 80-mm long by 60-mm high. Since the blast waves generated using exploding wires were cylindrical, the experiment was designed to be two dimensional. The entire model was confined between two parallel walls and the wire was placed horizontally outside the structure, 100 mm in front of the lower room entrance. The entrances to the two rooms spanned the whole width of the model in order to ensure that the two-dimensional nature of the event was preserved. Three Kistler 211B2 pressure transducers were flush mounted on the first floor wall, the second floor wall, and the second floor ceiling. The model was built from nonconductive plastic to prevent circuit short. The walls proximate to the wire were opaque to block the light emitted by the explosion while those inside the rooms were transparent to enable photography. The wire entrance holes were reinforced with steel to prevent damaging the plastic walls by the experiment.

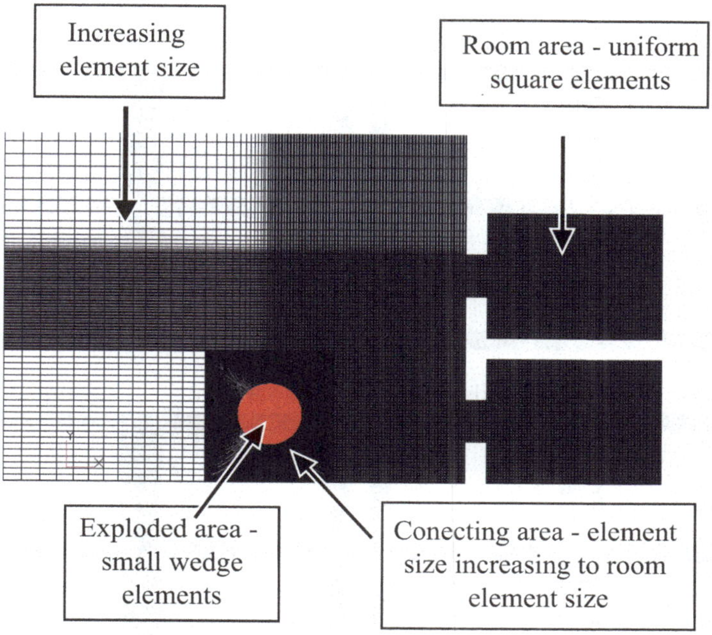

Fig. 9.9 Grid description of the solved numerical model

The grid of the numerical model shows that in the vicinity of the exploding region, the mesh in and between the rooms is gradually refined to permit high accuracy (see Fig. 9.9). The mesh inside the rooms was a 0.5-mm × 0.5-mm Cartesian grid while that outside the rooms was biased to allow element sizes to grow with the distance from the explosion area. Here also the explosion area was mimicked by placing a hot, dense cylindrical bubble in the exploded wire location.

9.4.1.3 Comparison of Experimental and Numerical Results

The experimental data acquired for the tested scenario enabled the examination and validation of a numerical model.

In contrast to full-scale experiments, in this laboratory scale test the implementation of a variety of accurate diagnostics yielded an amount of data that permitted performing a detailed comparison. The resulting visual images enabled a very good qualitative comparison to the numerical simulations, and the direct pressure measurements enabled a quantitative comparison. A series of experiments was performed using a 1.1-mm wire at 6 kV (the higher working point). Shadowgraph images taken in the experiment and shadowgraph representations obtained from the numerical solution are presented in the left and right panels of Fig. 9.10, respectively.

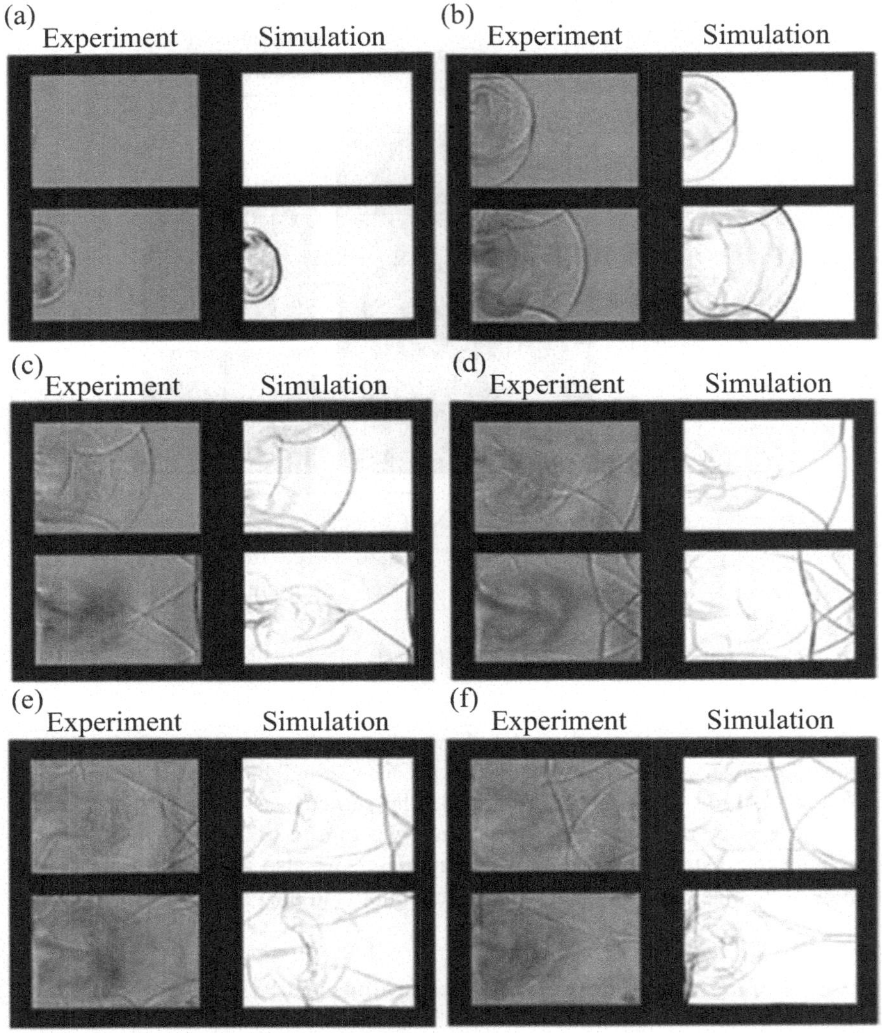

Fig. 9.10 Qualitative comparison between numerical (**a–f**, right panels) and experimental (**a–f**, left panels) results

The qualitative comparison shows that the results obtained from the numerical simulation were in good agreement with the experimental results. The incident blast waves that entered the rooms had similar characteristics in terms of, for example, their interaction with the structure and the intricate features of the different reflection phenomena. After a certain time, however, slight discrepancies emerged between the numerical and experimental results. Also, although the features of the reflections behind the incident blast wave were identical to those in front, their

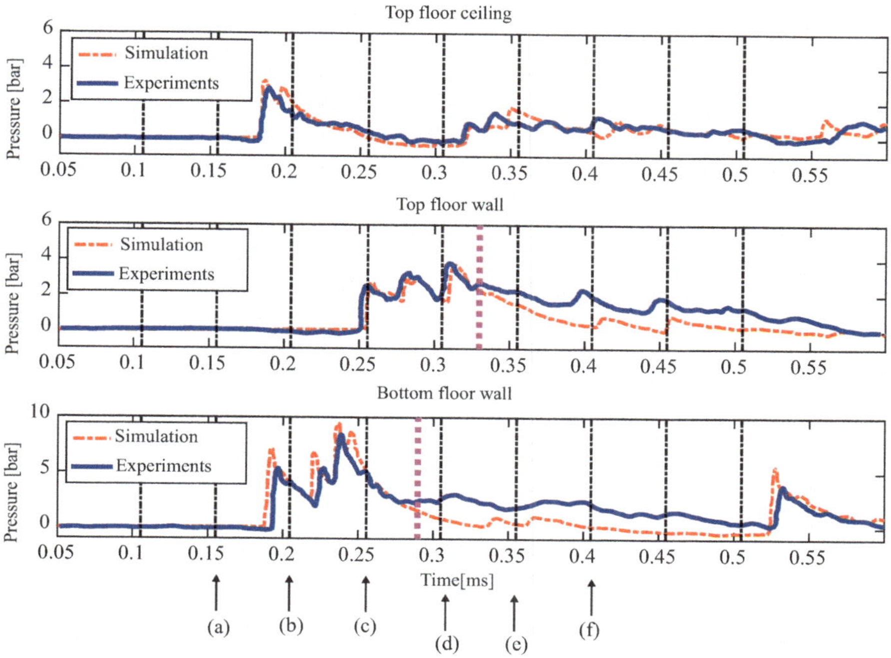

Fig. 9.11 Pressure comparisons between numerical and experimental results

locations were slightly different. A quantitative comparison of pressure history measurements to the corresponding results from the numerical simulation shows a discrepancy between the numerical and experimental results (Fig. 9.11). As the experiment progressed, the experimental pressure diverged from the calculated pressure at different times for the second floor wall and first floor wall measuring points (marked on the graphs with heavy dashed lines), but no such deviation was observed for the second floor ceiling.

9.4.1.4 Discussion and Conclusions

The foregoing described research demonstrated the use of an exploding wire technique to study blast wave–structure interaction. The experimental setup facilitated the implementation of sophisticated and sensitive diagnostics that permitted quantitative and qualitative comparisons with numerical tools. For the test case presented, it was shown that some discrepancies existed between the experimental and numerical results. A close examination of the results elicited a possible explanation for these discrepancies. An analysis of the captured images and the pressure history showed that the deviations between the two sets of results began after the interaction of the blast wave with the vortex created by the incoming flow.

It is clear that in its late stages, the interaction was not simulated correctly due to numerical limitations such as limited resolution, viscosity, and turbulence that were not accounted for, and three-dimensional effects. Furthermore, examining the pressure histories with the added information obtained from the images revealed that the waves traveling alongside the ceilings of both rooms were affected less by the incoming flow from the entrances. The pressure history on the second floor wall is in better agreement with the simulation due to this weaker interaction.

To apply this small-scale experimental system to full size structures, Kleine et al. (2005), Neuberger et al. (2007), Smith et al. (1992), and others have previously stated and demonstrated the ability to predict blast loads on full-scale structures based on the results of small-scale experiments. The common method of scaling a spherical explosion is the well-known Cranz-Hopkinson's "cube root" scaling law, which states that two similar geometries and charges that have different charge weights can be scaled using the cubic root of the charge energy. As shown in Baker (1973), the scaling parameters are:

$$\begin{aligned} Z &= \frac{R}{E^{1/3}}, \\ \tau^* &= \frac{\tau}{E^{1/3}}, \\ \zeta &= \frac{I}{E^{1/3}}, \end{aligned} \tag{9.3}$$

where Z is the scaled distance, τ^* is the scaled time, ζ is the scaled impulse, R is the radius from the explosion source, and E is the charge energy that can be replaced by the charge weight. Using these scaling parameters and a spherical rather than a cylindrical blast profile in the experiment, this scaling law can be used to scale up experimental results. Furthermore, using this scaling law, blast velocity, the pressure jump, and medium density are the same for the small- and full-scale setups, which leads to identical Mach and Euler numbers in both setups. These nondimensional numbers are highly relevant to this problem, where viscosity effects are negligible. In this case the two setups are similar and similitude is achieved.

9.4.2 *Case B: The Dependency of the Load Penetrating into a Structure on the Initial Conditions and the Internal Geometry*

Studies regarding shock and blast wave interaction with structures from a protective standpoint can be divided into two main groups. The first group includes studies that are more fundamental, where the impingement of a shock or blast is studied with regards to a specific element for purposes of either studying its response or to study the attenuation or enhancement properties of the element. The second group focuses on the load developing and the response of a structure subjected to an explosive event. While the first group usually deals with the physical mechanisms of shock structure interaction, developing flow fields and material response, the second group

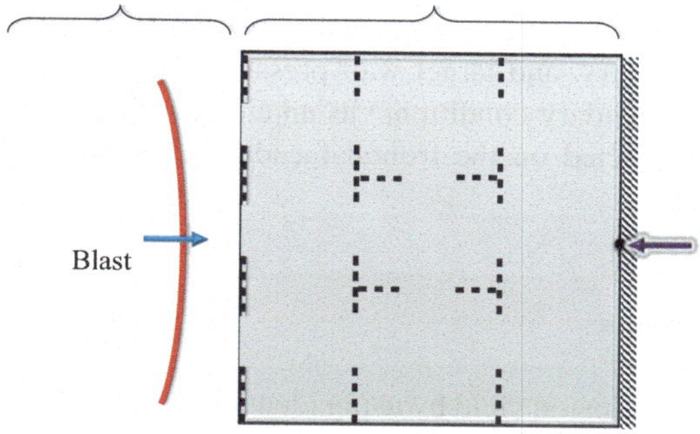

Fig. 9.12 The investigated problem. A blast wave impinging the frontal façade of a one-story building generates the pressure and impulse initial conditions. The blast propagating inside the structure is affected by the inner geometry. The point of interest is located at the center of the back wall

includes studies that are more typically focused on the application to structural design, load assessment, and survivability following an explosion event.

9.4.2.1 The Investigated Problem

This study aims to improve the understanding of the physical mechanisms determining the pressure buildup inside a structure following the impingement of an explosion-originated blast wave. Figure 9.12 broadly renders the important aspects of the studied problem. The study focuses on a single-story building exposed to a blast wave impinging on its frontal façade. The blast wave generates initial conditions on the frontal façade of the structure, namely pressure and impulse. Typically, following the initial impingement, a weaker blast wave enters the structure through the openings in the frontal façade, propagates through the structure, and reaches the back wall. Throughout its propagation, the initial wave diffracts and reflects off the internal walls. The diffracted waves reverberate throughout the structure, with some pressure exiting from the building façade openings, until eventually the reflections subside and the pressure returns to the atmospheric level. The main point of interest in this study was located along the structure's symmetry line, as shown in Fig. 9.12.

Resolving the three-dimensional diffraction pattern inside the complex structure is difficult and resource consuming. Furthermore, the wave propagation inside the structure and the resulting flow field highly depend on the internal geometry. Accurately resolving the internal flow field numerically for every change in the initial conditions or internal geometry would require a new simulation.

Rather than studying the detailed flow features and reflections, this study adopted a different approach to studying the pressure buildup dependency on the important parameters. We examined the scenario presented in Fig. 9.12 in terms of initial conditions, internal geometry, and target wall pressure. This approach considered the internal geometry (boundary conditions) as an element that modifies the initial pressure and impulse inflicted on the frontal façade and compares it to the load inflicted on the target wall.

9.4.2.2 Initial Conditions

Maintaining a constant explosion yield by using identical wires enables two possible ways to set the initial conditions imposed on the frontal façade of the structure (see Fig. 9.13). The first is by changing the frontal façade distance from the free air explosion. As the distance increases, the blast decays along with the imposed pressure and impulse. This method is effective but each distance generates a change in both the pressure profile and the impulse. A second method employs a stiff reflector placed around the explosion, which channels more energy towards the target. This method yields a different pressure—impulse combination than in the free air case.

The reflector permitted finding "semi-equivalent" initial conditions where the peak impulse generated at a specific distance from the free air explosion was the same as that generated with the reflector but at a different distance. For

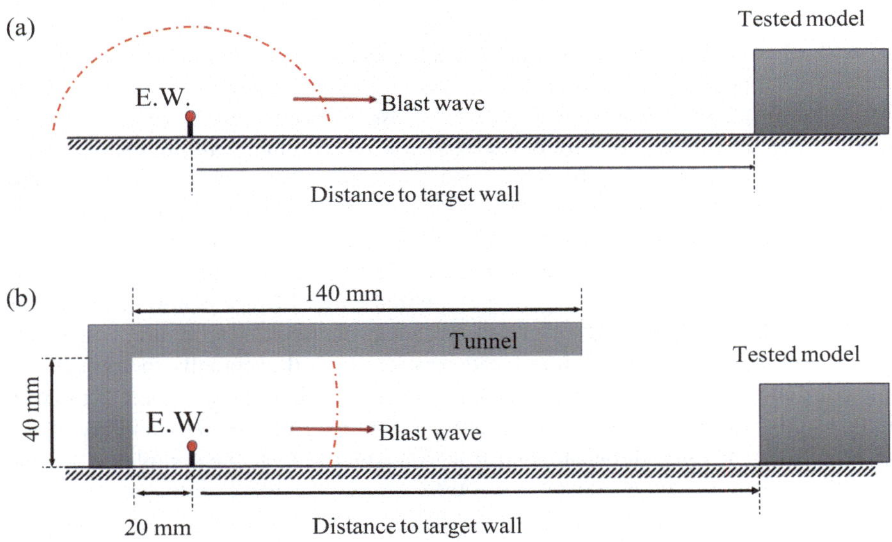

Fig. 9.13 The two methods chosen to generate the initial conditions: (**a**) Open-air ground-burst; (**b**) A reflector around the explosion area that channels more energy towards the structure

example, one combination used in this study was free air at 100 mm from the explosion and 300 mm from the explosion with the reflector. In these locations, the peak impulses were very similar though the peak overpressures were different. Though nonintuitive, this comparative method enabled the separation of one initial parameter: impulse or pressure.

The reflector was 40-mm high and 140-mm long. The gap between the inside back wall of the reflector and the exploding wire was 20 mm. All the experiments that were performed in this study were done using copper wires that were 70-mm long and 1.1 mm in diameter. The charged capacitor voltage was 6 kV for all of the experiments.

9.4.2.3 Repeatability

Explosions generated by the exploding wire technique are highly repeatable. The main parameters affecting the explosion strength are the wire material, wire dimensions, and the energy stored in the capacitor. To check the system repeatability, three experiments were conducted under the same conditions. The reflected pressure and impulse from these experiments measured at 150 mm from the explosion are presented in Fig. 9.14a, b, respectively. These results demonstrate the very good repeatability of the system and hence in the subsequent analysis each experiment appears only once.

9.4.2.4 Internal Geometry

Figure 9.15 depicts four different structure models used in this study. Each model had the same frontal façade spanning 92 mm by 25 mm (including external walls) with three 12 mm by 10 mm equally spaced windows. The models differed as follows:

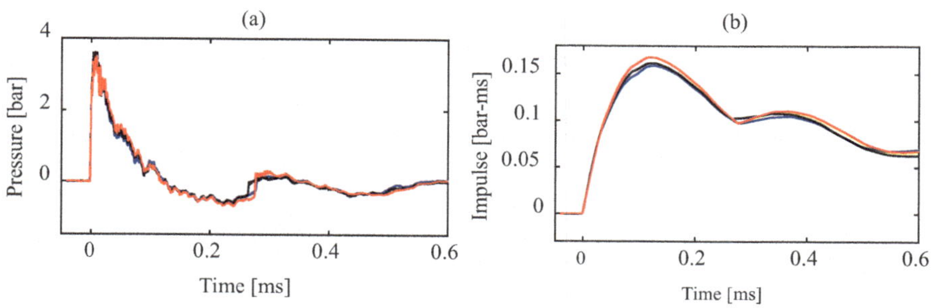

Fig. 9.14 Reflected pressure and impulse measured 150 mm from the explosion in three different experiments

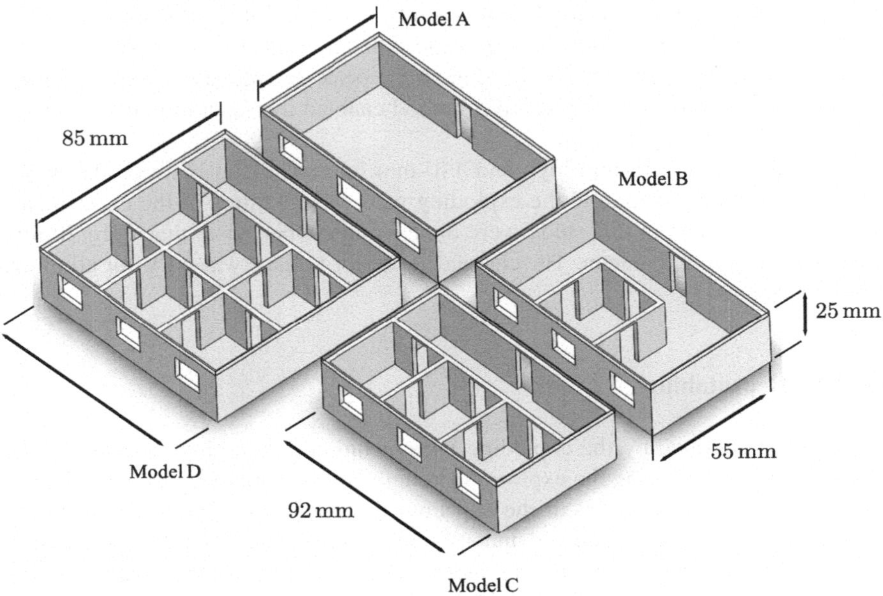

Fig. 9.15 Four one-story structure models ranging in size and in internal geometry

- Model A had no internal divisions.
- Model B had a 28 mm × 28 mm centered room partially obstructing the flow immediately behind the center window, in addition, an 8 mm × 20 mm doorway was located in the center of each internal wall.
- Model C had three internal rooms each 28 mm × 28 mm placed immediately behind the frontal façade wall. A doorway connected each pair of adjoining rooms and three more doorways were placed facing the façade and in symmetry with the three windows. All doors were 8 mm wide and 20 mm high.
- Model D was 85 mm long and had an additional row of internal rooms identical to the row of rooms in model C.

All the model parts were machined out of 2-mm thick polycarbonate sheets. The parts were then glued together assuring a tight seal. The models were positioned with the frontal façade directly facing the exploding wire at various distances. A pressure transducer (Kistler 211B3) was mounted on a wall adjacent to the target wall at the door opening (see schematic illustration in Fig. 9.16). The height and width of the mounting wall matched the structure so as not to influence the flow with the formation of external reflections.

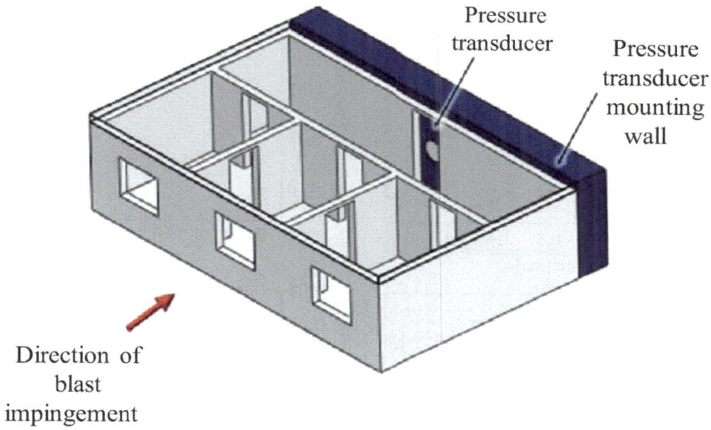

Fig. 9.16 Schematic description showing the pressure transducer mounted at the center of the opening in the target wall

9.4.2.5 Results

Two sets of experiments were performed in order to study the individual influence of the initial conditions and the internal geometry. In the first set, the internal geometry was varied while maintaining the initial conditions constant. In the second set, each model was subjected to varying initial conditions; a model was placed at a certain location, exposed to an open air explosion and then moved to a farther location with the reflector so that the imposed peak impulse was similar to the one measured in the closer open air experiments. The following sections aim to highlight the effects of each parameter.

Effects of the Internal Geometry

Figure 9.17 depicts the results obtained by exposing models A, B, and C to an open-air explosion with the model frontal façade placed at 150 mm from the explosion. Figure 9.17a, b contains the initial conditions; the measured pressure profile and impulse measured on the frontal façade before the effect of the internal geometry. These measurements were performed by mounting the pressure transducer at a distance of 150 mm from the explosion without the present of a model. Figure 9.17c, d presents the measured pressure and impulse profiles at the point of interest, i.e., the opening in the center of the target wall. To enable direct comparison between the results, all the presented recordings are shifted in time so that time $t = 0$ is when the incident wave reaches the transducer.

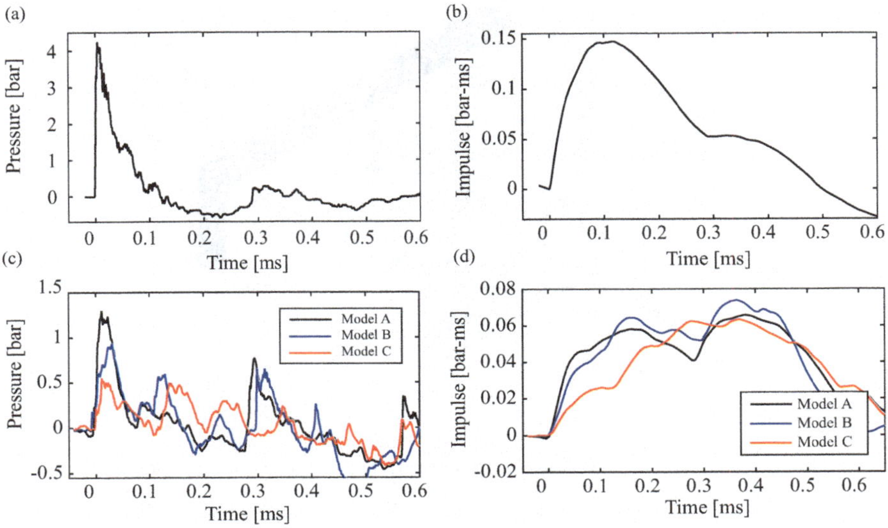

Fig. 9.17 Results from experiments performed with various internal geometries. (**a**, **b**) depict the reflected pressure and impulse developing at the frontal façade of the structures measured at 150 mm from the explosion without a model; (**c**, **d**) depict the pressure and impulse measured at the center of the target wall in each of the models A, B, and C

The results in Fig. 9.17 show that the overall measured impulse behavior at the target wall is similar in all of the experiments. As the models become more complex such as with model C we notice some changes in the impulse buildup, but still the overall behavior and peak impulses remain almost the same. Since each structure had the same windows at the frontal façade and the same initial conditions, the mass flux into the room and the incoming blast wave were also the same. The pressure profile, on the other hand, was highly influenced by the internal geometry. For example, Fig. 9.17c shows the reflection pattern through model A, a structure without internal divisions. The initial shock is measured arriving at the target wall and then reverberates between the frontal façade and the back wall. The arrival of this trapped shock produces definitive pressure jumps where there are only weak secondary reflections from the walls reaching the measurement point in between. In the experiments with model C, a significantly more complex structure, a pressure history featuring many shock reflections is visible. These weaker pressure jumps are accompanied by many minor reflections, which are probably transverse waves emanating from the interaction with internal walls.

A second effect revealed in Fig. 9.17c is that the initial pressure jump caused by the initial shock propagation strongly depends on the complexity of the inner geometry. The peak over pressure diminishes with exceedingly complex geometries. Subsequent reflections seem to be more similar in strength between the models.

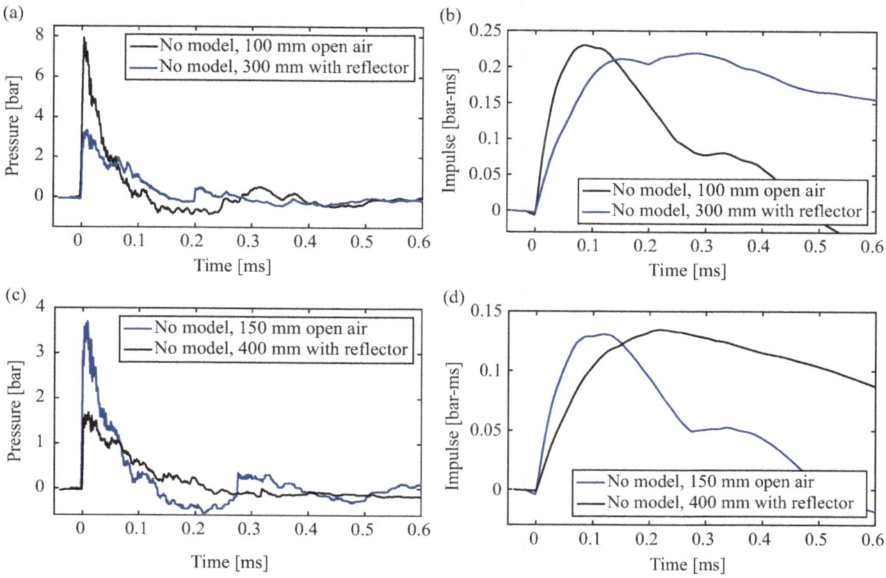

Fig. 9.18 Comparison of the pressure-impulse combinations that were obtained in open air and with reflector explosions. The explosion with reflector provides a peak impulse at a certain location that matches that of an open-air explosion at a closer location

Effects of the Initial Conditions

A second set of experiments was performed to study the effects of the initial conditions on the pressure buildup at the target wall. As stated previously, changing the distance between the exploding wire and the model frontal façade alters the initial conditions. The impulse and pressure change in different manners, and hence this method offers no comparable data. An alternative method overcomes this limitation. A reflector, shown in Fig. 9.13b, focuses the blast energy thereby providing at a certain location a peak impulse that matches that of an open-air explosion at a closer range. Figure 9.18 presents two combinations of open-air and reflector explosions. It is impossible to generate the same impulse history at the frontal façade, but Fig. 9.18 depicts two combinations that generate impulses sufficiently similar for our purposes, especially when the peak impulse is concerned.

The two combinations shown in Fig. 9.18 generate similar *peak impulses* on the structure while imposing very different pressure loads both in peak over pressure and in the duration. It is important to note that the peak impulse is reached when the positive pressure duration ends, i.e., the pressure returns to atmospheric conditions. Since the open-air explosion generates an additional negative pressure phase, the impulse declines much faster. This effect, however, has little influence on the conditions inside the structure since low pressure propagates inwards at a slower speed. The mass flow into the structure effectively stops when the impulse begins to decline.

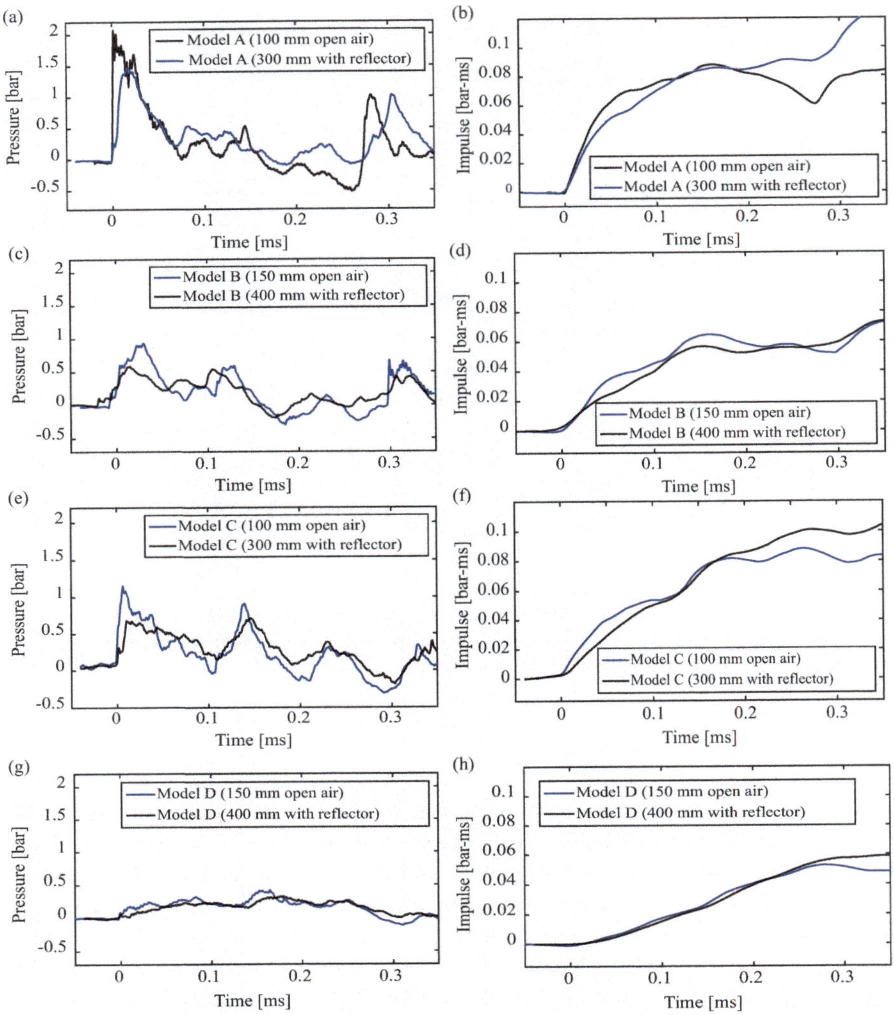

Fig. 9.19 Experimental results depicting the target wall pressure and impulse profiles in the various models. The experiments were performed with different initial conditions in which the inflicted peak impulses were kept similar

The initial conditions shown in Fig. 9.18 were inflicted on the models shown in Fig. 9.15 by placing the façade at the same location where they were measured (the pressure transducer mounting wall was placed at different locations accordingly). Figure 9.19 depicts the over pressures and impulses measured at the end wall of models A, B, C, and D. A single set of initial conditions is presented for each model.

As expected, since the two impulse initial conditions were similar up to the peak, the impulses on the target wall are also in a reasonable agreement. In the open-air experiments, the low pressure propagates into the structure and lowers the impulse after the peak as seen in the initial conditions (Fig. 9.18).

The pressure measurements recorded at the target walls, on the other hand, are not so intuitive. Despite very different initial pressures inflicted on the façade for the open air and reflector explosions (up to twice the peak over pressure), the pressure profiles inside the structure were remarkably similar. In fact, the pressure profiles display similar over pressure as well as similar reflection patterns. To explain this, we refer to the results shown in Fig. 9.17. We saw that imposing the same initial conditions in terms of both pressure and impulse caused the impulses at the target wall to match but with very different reflection patterns. In this case, the dominance of the internal geometry is even more pronounced since the reflection patterns are not only highly affected by the internal geometry but also impede the propagation of stronger shocks inside the structure. To support this notion, a comparison between two different internal geometries shows that as the complexity increases the pressure profiles become even more similar.

9.4.2.6 Application to Large Explosion Modeling

The above findings have valuable implications for simulating large explosion in laboratory settings. Though very repeatable and easy to employ, the exploding wire method generates weak explosions. A previous study showed that at the system used here, i.e., exploding a 70-mm long, 1.1-mm diameter copper wire with 6 kV in the capacitor generates an energy equivalent to the explosion of 0.1 g of TNT (Ram and Sadot 2012). Consequently, simulating large explosions in urban scenarios, such as one ton of TNT, would necessitate scaling down to miniscule proportions. The results here provide a tool to address this problem.

To demonstrate the implications of the results shown above, a full-scale simulation of an explosion event at which an 800 kg TNT charge explodes 15 m from the frontal façade of a structure is evaluated. Figure 9.20 depicts the overall dimensions of the full-scale scenario. The structure has three equally spaced 1 m × 1 m windows at its frontal façade, and two 1 m × 2 m windows at its sides. Inside, a 3 m × 3 m room was located adjacent to the frontal façade, behind the center window, partially obstructing the path to an opening at the target wall.

Fully scaling down the model, by means of the Cranz-Hopkinson law, to meet the exploding wire system capabilities requires an impracticably small model (in the order of 1:1000). The model would be so small that the pressure transducer could not fit the opening at the target wall.

In light of this restriction, we chose a more feasible small-scale ratio of 1:100 to simulate the explosion by means of the exploding wire system. Properly simulating the explosion would further require creating a scaled down 0.8 g hemispherical explosion at a distance of 150 mm from the frontal façade of a scaled down model. Since we could not satisfy these conditions under the exploding wire system limitations, we used the reflector and found the location where the produced impulse matched a 0.8 g TNT explosion at a 150 mm range. The desired pressures and

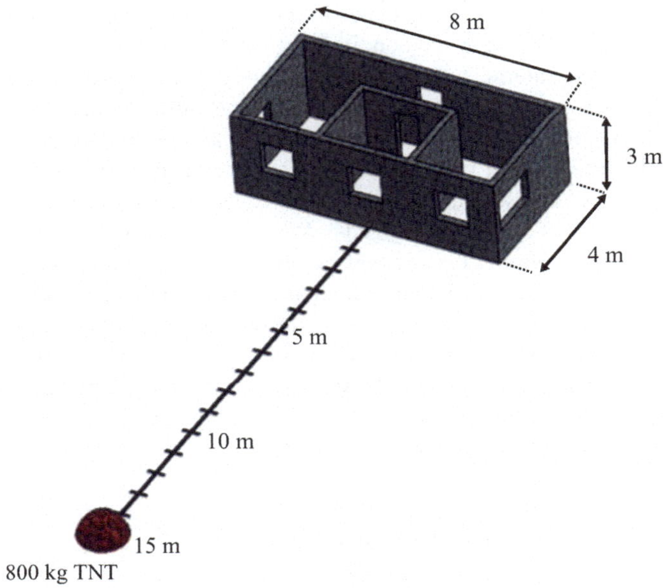

Fig. 9.20 The full-scale scenario in which an 800 kg hemispherical TNT charge explodes 15 m from a one-story structure façade. (The roof is transparent)

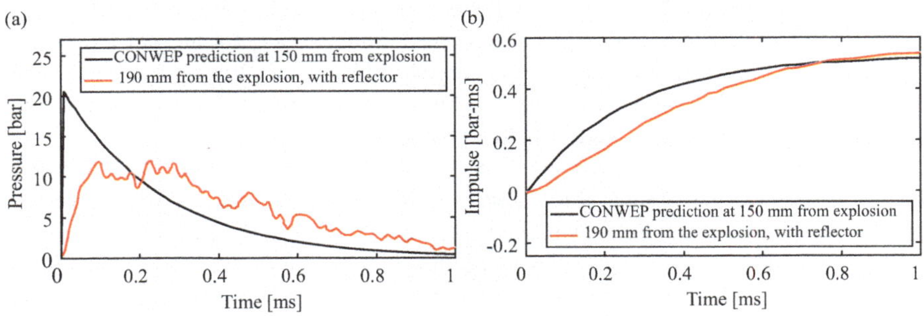

Fig. 9.21 Pressure and impulse profiles required to simulate an 800 kg explosion in a 1:100 scale. The desired conditions are that of a 0.8 g TNT charge detonated 150 mm from the model. The closest conditions were found using the reflector at 190 mm from the exploding wire

impulses were obtained by means of the CONWEP software (1992). Figure 9.21 depicts the closest match that was found, at a distance of 190 mm from the exploding wire. As concluded in Sect. 9.4.2.5 we selected the distance that provided the most similar impulse.

Numerical simulations of the full-scale model were performed using MSC/Dytran commercial software. The numerical model was validated with full scale experiments and was found to be in good agreement (Ostraich et al. 2009)

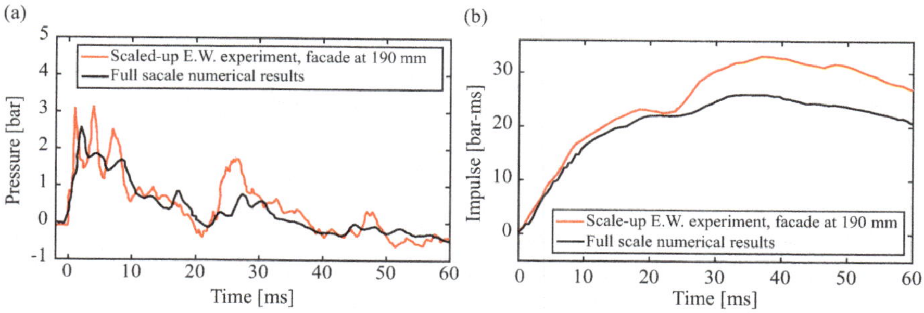

Fig. 9.22 A comparison between a numerical simulation and an exploding wire experiment at the target wall. While the initial imposed conditions on the frontal façade only matched in terms of impulse. A very good resemblance was found in both the impulse and the pressure at the target wall

with them. In order to simplify the numerical model it was assumed that during the initial wave propagation inside the structure the walls were non-deformable. A symmetry assumption was also used and only half of the structure was simulated. The explosion was modeled as a hot air bubble burst calibrated to yield the appropriate TNT equivalent (Ram and Sadot 2012). The blast loading on the building frontal façade was validated by CONWEP predictions, which also yielded satisfactory agreement.

The small-scale model was built from polycarbonate sheets in the same manner as the previous models were made and placed in the exploding wire system. Figure 9.22 shows a comparison between the experimental results and numerical simulation of the full-scale experiments. To display the comparison, the pressure and the impulse histories from the small-scale experiments were scaled up according to the Cranz-Hopkinson law to the full-scale simulation.

The small-scale experiment and the numerical solution agree in both impulse and pressure. It could be expected that by tailoring the impulse imposed on the façade we should get the proper impulse at the target wall. However, an agreement between the pressure profiles was not initially expected but could be explained based on the results presented in the previous section. The impulse transfer through the structure along with the internal geometry assured that the pressure profile was very similar to the one recorded in the simulation regardless of the incorrect pressure imposed on the frontal façade.

9.4.3 Conclusions

The current study empirically explored the effects of initial conditions and internal geometry on the pressure developing inside a single-story building. It was found that when inflicting similar impulses at the frontal façade of the structure, the internal geometry determines the pressure profile that develops at the target wall regardless of the initial pressure profile.

These inferences were applied to the experimental study of a blast event in a small-scale setup. The presented findings imply that in the case where the load developed in an interior of a complex structure by exterior blast event is of interest, it is satisfactory to impose the correct impulse at the façade to acquire adequate results in terms of both impulse and pressure inside the structure.

Furthermore, as the complexity of the structure increases, the larger number of reflections facilitated by the internal divisions contribute to the independency of the target wall loading in the imposed pressure profile. The results show the consistency of this phenomenon and will in the future enable faster examination of protective structures in laboratory settings.

9.5 Final Remarks

The experimental system presented in this chapter is capable of producing small-scale, well-controlled explosions in a laboratory setting. The available diagnostic system in use provides valuable insight which enables better understanding of the physics of the interaction of cylindrical and spherical blast waves with structures.

The importance of experimentally validating numerical simulations was presented. Its ease of use and low operational costs together with the sophisticated diagnostics available facilitate accurate experimental results that can highlight the strengths and weaknesses of the numerical models for any specific problems of blast–structure interaction. After construction, calibration, and testing of the experimental system, various scenarios can be studied and numerical codes can be verified and validated.

To create three-dimensional blast waves, a small knot made from a copper wire was used. This setup created spherical blast waves that led to more realistic scenarios. The spherical nature of the exploding knot was tested and proved. The pressure histories recorded in a spherical blast experiment were compared to CONWEP generated results and found to be in good agreement.

Using dimensional analysis and similitude, experimental prediction of the load induced by the blast wave in this small-scale experimental setup can be used to assess the expected dynamic loads in the full-scale setup.

Acknowledgments The authors would like to thank Dr. A. L. Levin for his constructive remarks. This study is partially supported by the Israel Ministry of Defense grants numbers 4440507635 and 4440130927. O. Ram is supported by the Adams scholarship program of the Israeli Academy of Science.

References

Baker, W. (1973). *Explosions in air* (pp. 54–77). Austin: University of Texas Press.

Bennet, F. D. (1959). Flow fields produced by exploding wires. In W. G. Moore & H. K. Chace (Eds.), *Exploding wires proceedings* (Vol. 1, pp. 211–266). New York: Plenum Press.

Buntzen, R. R. (1962). The use of exploding wires in the study of small-scale underwater explosions. In W. G. Moore & H. K. Chace (Eds.), *Exploding wires proceedings* (Vol. 2, pp. 195–205). New York: Plenum Press.

Chase, W. G., & Moore, H. K. (1959). *Exploding wires*. New York: Plenum Press. 4 Volumes.

CONWEP. (1992). Conventional weapons effects, U. S. Army Engineer Waterways Experiment Station, CEWES-SS-R, 20 Aug 1992.

Cheval, K., Loiseau, O., & Vala, V. (2010). Laboratory scale tests for the assessment of solid explosive blast effects. Part I: Free-field test campaign. *Journal of Loss Prevention in the Process Industries, 23*(5), 613–621.

Dewey, J. M., McMillin, D. J., & Classen, D. F. (1977). Photogrammetry of spherical shocks reflected from real and ideal surfaces. *Journal of Fluid Mechanics, 81*, 701–717.

Dewey, J. M. (2001). Expending spherical shocks. In G. Ben-Dor, O. Igra, & T. Elperin (Eds.), *Handbook of shock waves* (Vol. 2, pp. 441–481). San Diego: Academic Press.

Hargather, J., & Settles, G. S. (2007). Optical measurement and scaling of blasts from gram-range explosive charges. *Shock Waves, 17*, 215–223.

Higashino, F., Henderson, L. F., & Shimizu, F. (1991). Experiments on the interaction of a pair of cylindrical weak blast waves in air. *Shock Waves, 1*, 275–284.

Jiang, Z., Takayama, K., Moosad, K. P. B., Onodera, O., & Sun, M. (1998). Numerical and experimental study of a micro-blast wave generated by pulsed-laser beam focusing. *Shock Waves, 8*, 337–349.

Kleine, H., & Takayama, K. (2004). Laboratory-scale blast wave phenomena. In B. Milton, T. Saito, & M. Sun (Eds.), *Proceedings of the symposium on Interdisciplinary Shock Wave Research* (pp. 257–276). Tohoku University.

Kleine, H., Timofeev, E., & Takayama, K. (2005). Laboratory-scale blast wave phenomena – Optical diagnostics and applications. *Shock Waves, 14*, 343–357.

Lin, S. C. (1954). Cylindrical shock waves produced by instantaneous energy release. *Journal of Applied Physics, 25*, 54–57.

Needham, C. E. (2010). *Blast waves*. Heidelberg: Springer.

Neuberger, A., Peles, S., & Rittel, D. (2007). Scaling the response of circular plates subjected to large and close-range spherical explosions. Part I: Air-blast loading. *International Journal of Impact Engineering, 34*, 859–873.

Ostraich, B., Kivity, Y., Anteby, I., Sadot, O., & Ben-Dor, G. (2009). Load assesment on safe rooms doors report. Shock Tube Laboratory, Protective Technologies R&D Center, Faculty of Engineering Sciences, Ben-Gurion University of the Negev, Beer-Sheva, Israel (in Hebrew).

Ram, O., & Sadot, O. (2012). Implementation of the exploding wire technique to study blast-wave–structure interaction. *Experiments in Fluids, 53*, 1335–1345. https://doi.org/10.1007/s00348-012-1339-8.

Reichenbach, H., & Neuwald, P. (2000). *Fluid-dynamics of explosions in multi-chamber systems phenomenology test program*. Freiburg: Ernst-Mach Institute.

Reithel, R. J., Blackburn, J. H., Seay, G. E., & Skolnick, S. (1962). The current pause in an exploding wire. In W. G. Moore & H. K. Chace (Eds.), *Exploding wires proceedings* (Vol. 1, pp. 19–32). New York: Plenum Press.

Rose, T. A., & Smith, P. D. (2002). Influence of the principal geometrical parameters of straight city streets on positive and negative phase blast wave impulses. *International Journal of Impact Engineering, 27*, 239–376.

Settles, G. S. (2001). *Schlieren and shadowgraph techniques: Visualizing phenomena in transparent media*. New York: Springer.

Smith, P. D., Mays, G. C., Rose, T. A., Teo, K. G., & Roberts, B. J. (1992). Small scale models of complex geometry for blast over pressure assessment. *International Journal of Impact Engineering, 12*, 345–360.

Smith, P. D., & Hetherington, J. G. (1994). *Blast and ballistic loading of structures* (pp. 145–223). Oxford: Butterworth-Heinemann.

Taylor, G. I. (1950). The formation of a blast wave by a very intense explosion. I. Theoretical discussion. *Proceedings of the Royal Society of London. Series A, Mathematical and Physical Sciences, 201*(1065), 159–174.

Zyskowski, A., Sochet, I., Mavrot, G., Bailly, P., & Renard, J. (2004). Study of the explosion process in a small scale experiment – Structural loading. *Journal of Loss Prevention, 17*, 291–299.

Appendix A
Parameters Values for Fitted Curves: Gaseous Mixtures—Hemispherical Charges

See Tables A.1, A.2, A.3, A.4, A.5, A.6, A.7, A.8, A.9, A.10, A.11, A.12, A.13, A.14, A.15, and A.16.

Table A.1 Hemispherical gaseous charge—overpressure versus scaled distance in energy

$\text{Ln}\ \frac{\Delta P}{P_0} = A_0 + \sum_{i=1}^{i=n} A_i\ (\text{Ln}\ \lambda)^i$

λ (m MJ$^{-1/3}$)

Authors	A_0	A_1	A_2	Range λ
Prisme	0.574237	−1.77465	0.0945742	$0.4 \leq \lambda \leq 25.0$
Fraunhoffer	0.821525	−1.68252	0.0662674	$1.6 \leq \lambda \leq 28.4$
Dorofeev	0.546206	−1.93552	0.169384	$0.5 \leq \lambda \leq 10.3$
Brossard	0.336072	−1.56643	0.0647514	$0.9 \leq \lambda \leq 23.2$
Dewey	0.576661	−2.06142	0.215509	$0.4 \leq \lambda \leq 23.7$

Table A.2 Hemispherical gaseous charge—arrival time versus scaled distance in energy

$\text{Ln}\ T_a.E^{-1/3} = A_0 + \sum_{i=1}^{i=n} A_i\ (\text{Ln}\ \lambda)^i$

T_a (ms) E (MJ) λ (m MJ$^{-1/3}$)

Authors	A_0	A_1	A_2	A_3	A_4	Range λ
Prisme	0.0825981	1.85804	−0.030773	−0.168237	0.0386282	$0.4 \leq \lambda \leq 25.0$
Fraunhoffer	0.0997063	1.56703	−0.0882913			$1.6 \leq \lambda \leq 28.4$
Dewey	1.13084	1.02787	−0.0238925			$0.4 \leq \lambda \leq 23.7$

I. Sochet (ed.), *Blast Effects*, Shock Wave and High Pressure Phenomena,
https://doi.org/10.1007/978-3-319-70831-7

Table A.3 Hemispherical gaseous charge—positive impulse versus scaled distance in energy

$\text{Ln}\, I^+.E^{-1/3} = A_0 + \sum_{i=1}^{i=n} A_i\, (\text{Ln}\, \lambda)^i$

I^+ (bar ms) E (MJ) λ (m MJ$^{-1/3}$)

Authors	A_0	A_1	A_2	A_3	Range λ
Prisme	−0.77627	−0.738704	−0.244664	0.0497853	$0.4 \leq \lambda \leq 25.0$
Fraunhoffer	−0.37508	−0.861535	−0.038644		$1.6 \leq \lambda \leq 28.4$
Dorofeev	−0.295739	−0.968			$0.5 \leq \lambda \leq 10.3$
Brossard	−0.112242	−4.4694	7.02388	−7.07077	$0.9 \leq \lambda \leq 23.2$
		A_4	A_5	A_6	
		3.68523	−0.940872	0.0924594	

Table A.4 Hemispherical gaseous charge—positive phase duration versus scaled distance in energy

$\text{Ln}\, T^+.\lambda^{-1/3} = A_0 + \sum_{i=1}^{i=n} A_i\, (\text{Ln}\, \lambda)^i$

T^+ (ms) E (MJ) λ (m MJ$^{-1/3}$)

Authors	A_0	A_1	A_2	A_3	Range λ
Prisme	0.121706	0.449697	−0.0243137	−0.00966258	$0.4 \leq \lambda \leq 25.0$
Fraunhoffer	0.113682	−0.104311	0.960696	−0.628328	$1.6 \leq \lambda \leq 28.4$
			A_4	A_5	
			0.177585	−0.0192233	
Brossard	−0.209479	1.68576	−3.6275	4.71075	$0.9 \leq \lambda \leq 23.2$
		A_4	A_5	A_6	
		−3.05501	0.954646	−0.114651	

Table A.5 Hemispherical gaseous charge—scaled distance in energy versus overpressure

$\text{Ln}\, \lambda = A_0 + \sum_{i=1}^{i=n} A_i\, (\text{Ln}\, \frac{\Delta P}{P_0})^i$

λ (m MJ$^{-1/3}$)

Authors	A_0	A_1	A_2	Range $\frac{\Delta P}{P_0}$
Prisme	0.326301	−0.580746	0.0249897	$0.015 \leq \frac{\Delta P}{P_0} \leq 13.400$
Fraunhoffer	0.510627	−0.659578	0.0003029	$0.015 \leq \frac{\Delta P}{P_0} \leq 1.000$
Dorofeev	0.271555	−0.545126	0.0413703	$0.04 \leq \frac{\Delta P}{P_0} \leq 7.30$
Brossard	0.222637	−0.638139	0.024601	$0.02 \leq \frac{\Delta P}{P_0} \leq 10.70$
Dewey	0.249097	−0.540493	0.0494433	$0.02 \leq \frac{\Delta P}{P_0} \leq 1.35$

Table A.6 Hemispherical gaseous charge—scaled distance in energy versus arrival time

$\text{Ln}\,\lambda = A_0 + \sum_{i=1}^{i=n} A_i\,(\text{Ln}\,T_a.E^{-1/3})^i$

λ (m MJ$^{-1/3}$) T_a (ms) E (MJ)

Authors	A_0	A_1	A_2	Range $T_a.E^{-1/3}$
Prisme	−0.0547876	0.850326	0.0629026	$0.2 \leq T_a.E^{-1/3} \leq 66.0$
Fraunhoffer	−0.0088649	0.562194	0.0470502	$2.5 \leq T_a.E^{-1/3} \leq 80.3$
Dewey	−1.07542	0.940162	0.0181726	$0.003 \leq T_a.E^{-1/3} \leq 43.000$

Table A.7 Hemispherical gaseous charge—scaled distance in energy versus positive impulse

$\text{Ln}\,\lambda = A + \sum_{i=1}^{i=n} A_i\,(\text{Ln}\,I^+.E^{-1/3})^i$

λ (m MJ$^{-1/3}$) I^+ (bar ms) E (MJ)

Authors	A_0	A_1	A_2	A_3	Range $I^+.E^{-1/3}$
Prisme	−1.04889	−1.47713	−0.206919	−0.0240238	$0.57 \leq I^+.E^{-1/3} \leq 16.10$
Fraunhoffer	−0.281433	−0.922103	0.0936892	0.0227545	$0.03 \leq I^+.E^{-1/3} \leq 0.40$
Dorofeev	−0.305515	−1.03306			$0.07 \leq I^+.E^{-1/3} \leq 1.26$
Brossard	0.0767023	0.343772	0.615244	0.0857423	$0.016 \leq I^+.E^{-1/3} \leq 0.0356$

Table A.8 Hemispherical gaseous charge—scaled distance in energy versus positive phase duration

$\text{Ln}\,\lambda = A + \sum_{i=1}^{i=n} A_i\,(\text{Ln}\,T^+.E^{-1/3})^i$

λ (m MJ$^{-1/3}$) T^+ (ms) E (MJ)

Authors	A_0	A_1	A_2	A_3	A_4	Range $T^+.E^{-1/3}$
Prisme	−0.150191	1.45196	1.39913			$0.3 \leq T^+.E^{-1/3} \leq 2.7$
Fraunhoffer	−0.945785	9.8672	−20.147	19.8663	−6.41636	$1.3 \leq T^+.E^{-1/3} \leq 3.3$
Brossard	0.362614	1.96402	0.730464			$0.7 \leq T^+.E^{-1/3} \leq 1.9$

Table A.9 Hemispherical gaseous charge—overpressure versus scaled distance in mass

$\text{Ln}\,\frac{\Delta P}{P_0} = A_0 + A_i\,(\text{Ln}\,Z)^i$

Z (m kg$^{-1/3}$)

Authors	A_0	A_1	A_2	Range Z
Prisme	1.98458	−1.93917	0.100553	$0.8 \leq Z \leq 53.4$
Fraunhoffer	1.44779	−1.73115	0.0662674	$2.3 \leq Z \leq 41.0$
Dorofeev	1.12408	−1.88654	0.12445	$0.8 \leq Z \leq 14.7$
Brossard	0.89644	−1.61209	0.0647514	$1.3 \leq Z \leq 33.1$
Dewey	2.2055	−2.37772	0.215509	$0.9 \leq Z \leq 49.5$

Table A.10 Hemispherical gaseous charge—arrival time versus scaled distance in mass

$\text{Ln}\ T_a.M^{-1/3} = A_0 + \sum_{i=1}^{i=n} A_i\ (\text{Ln Z})^i$

T_a (ms) M (kg) Z (m kg$^{-1/3}$)

Authors	A_0	A_1	A_2	Range Z
Prisme	−0.588911	1.87672	−0.125803	$0.8 \leq Z \leq 53.4$
Fraunhoffer	−0.786667	1.87751	−0.129502	$2.3 \leq Z \leq 41.0$
Dewey	1.09752	1.06293	−0.0238926	$0.006 \leq Z \leq 30.500$

Table A.11 Hemispherical gaseous charge—positive impulse versus scaled distance in mass

$\text{Ln}\ I^+.M^{-1/3} = A_0 + \sum_{i=1}^{i=n} A_i\ (\text{Ln Z})^i$

I^+ (bar ms) M (kg) Z (m kg$^{-1/3}$)

Authors	A_0	A_1	A_2	A_3	Range Z
Prisme	0.392073	−0.274874	−0.359356	0.0497853	$0.8 \leq Z \leq 53.4$
Fraunhoffer	0.29406	−0.670527	−0.346769	0.209148	$2.3 \leq Z \leq 41.0$
			A_4	A_5	
			−0.0593357	0.00597998	
Dorofeev	0.402815	−0.968			$0.8 \leq Z \leq 14.7$
Brossard	−0.168559	−0.913777	0.0184993	−0.0131382	$1.3 \leq Z \leq 33.1$

Table A.12 Hemispherical gaseous charge—positive phase duration versus scaled distance in mass

$\text{Ln}\ T^+.M^{-1/3} = A_0 + \sum_{i=1}^{i=n} A_i\ (\text{Ln Z})^i$

T^+ (ms) M (kg) Z (m kg$^{-1/3}$)

Authors	A_0	A_1	A_2	A_3	Range Z
Prisme	0.534325	0.469945	−0.00205389	−0.00966258	$0.8 \leq Z \leq 53.4$
Fraunhoffer	0.682562	−1.09981	1.80518	−0.914797	$2.3 \leq Z \leq 41.0$
			A_4	A_5	
			0.212841	−0.0192223	
Brossard	−0.105709	0.778628	−0.121826		$1.3 \leq Z \leq 33.1$

Table A.13 Hemispherical gaseous charge—scaled distance in mass versus overpressure

$\text{Ln Z} = A_0 + \sum_{i=1}^{i=n} A_i\ (\text{Ln}\ \frac{\Delta P}{P_0})^i$

Z (m kg$^{-1/3}$)

Authors	A_0	A_1	A_2	Range $\frac{\Delta P}{P_0}$
Prisme	1.07962	−0.578252	0.0266957	$0.015 \leq \frac{\Delta P}{P_0} \leq 13.400$
Fraunhoffer	0.877546	−0.659578	0.0003029	$0.015 \leq \frac{\Delta P}{P_0} \leq 1.000$
Dorofeev	0.612653	−0.583349	0.029536	$0.04 \leq \frac{\Delta P}{P_0} \leq 7.30$
Brossard	0.575234	−0.638139	0.024601	$0.02 \leq \frac{\Delta P}{P_0} \leq 1.35$
Dewey	0.982952	−0.540493	0.0494433	$0.02 \leq \frac{\Delta P}{P_0} \leq 10.7$

Table A.14 Hemispherical gaseous charge—scaled distance in mass versus arrival time

$\text{Ln}\, Z = A_0 + \sum_{i=1}^{i=n} A_i\, (\text{Ln}\, T_a.M^{-1/3})^i$

$Z\,(\text{m kg}^{-1/3})$ T_a (ms) M (kg)

Authors	A_0	A_1	A_2	Range $T_a.M^{-1/3}$
Prisme	0.343366	0.49843	0.0470081	$0.4 \leq T_a.M^{-1/3} \leq 141.8$
Fraunhoffer	0.456797	0.505484	0.0510854	$3.5 \leq T_a.M^{-1/3} \leq 41.0$
Dewey	−1.02172	0.91349	0.0181726	$0.006 \leq T_a.M^{-1/3} \leq 89.500$

Table A.15 Hemispherical gaseous charge—scaled distance in mass versus positive impulse

$\text{Ln}\, Z = A_0 + \sum_{i=1}^{i=n} A_i\, (\text{Ln}\, I^+.M^{-1/3})^i$

$Z\,(\text{m kg}^{-1/3})$ I^+ (bar ms) M (kg)

Authors	A_0	A_1	A_2	Range $I^+.M^{-1/3}$
Prisme	0.741899	−1.10717	−0.050778	$3.4 \leq I^+.M^{-1/3} \leq 6.9$
Fraunhoffer	0.347769	−1.18704	−0.0586965	$0.60 \leq I^+.M^{-1/3} \leq 0.04$
Dorofeev	0.416131	−1.03306		$0.014 \leq I^+.M^{-1/3} \leq 0.230$
Brossard	−0.04302	−1.00385	−0.00868037	$0.02 \leq I^+.M^{-1/3} \leq 0.70$

Table A.16 Hemispherical gaseous charge—scaled distance in mass versus positive phase duration

$\text{Ln}\, Z = A_0 + \sum_{i=1}^{i=n} A_i\, (\text{Ln}\, T^+.M^{-1/3})^i$

$Z\,(\text{m kg}^{-1/3})$ T^+ (ms) M (kg)

Authors	A_0	A_1	A_2	Range $T^+.M^{-1/3}$
Prisme	0.327782	−0.696835	1.39913	$1.4 \leq T^+.M^{-1/3} \leq 5.9$
Fraunhoffer	0.633227	−0.324146	1.2839	$1.8 \leq T^+.M^{-1/3} \leq 4.7$
Brossard	0.113517	1.4489	0.730464	$1.0 \leq T^+.M^{-1/3} \leq 3.8$

Appendix B
Parameters Values for Fitted Curves: Gaseous Mixtures—Spherical Charges

See Tables B.1, B.2, B.3, B.4, B.5, B.6, B.7, and B.8.

Table B.1 Spherical gaseous charge—overpressure versus scaled distance in energy

$$\mathrm{Ln}\,\frac{\Delta P}{P_0} = A_0 + \sum_{i=1}^{i=n} A_i\,(\mathrm{Ln}\,\lambda)^i$$

λ (m MJ$^{-1/3}$)

Authors	A_0	A_1	A_2	Range λ
Prisme	0.36042	−1.87327	0.152015	$0.7 \le \lambda \le 6.7$
Fraunhoffer	1.06533	−1.3693	−0.031925	$1.4 \le \lambda \le 31.3$
Brossard	0.864309	−1.75127	0.121468	$0.48 \le \lambda \le 26.60$

Table B.2 Spherical gaseous charge—arrival time versus scaled distance in energy

$$\mathrm{Ln}\,T_a.E^{-1/3} = A_0 + \sum_{i=1}^{i=n} A_i\,(\mathrm{Ln}\,\lambda)^i$$

T_a (ms) E (MJ) λ (m MJ$^{-1/3}$)

Authors	A_0	A_1	A_2	Range λ
Prisme	0.221213	1.56146	−0.112726	$0.7 \le \lambda \le 6.7$
Fraunhoffer	0.334677	1.37613	−0.0527822	$1.4 \le \lambda \le 31.3$

I. Sochet (ed.), *Blast Effects*, Shock Wave and High Pressure Phenomena,
https://doi.org/10.1007/978-3-319-70831-7

Table B.3 Spherical gaseous charge—scaled distance in energy versus overpressure

$\text{Ln}\,\lambda = A_0 + \sum_{i=1}^{i=n} A_i\,(\text{Ln}\,\frac{\Delta P}{P_0})^i$				
λ (m MJ$^{-1/3}$)				
Authors	A_0	A_1	A_2	Range $\frac{\Delta P}{P_0}$
Prisme	0.188042	−0.548621	0.0360489	$0.07 \leq \frac{\Delta P}{P_0} \leq 2.40$
Fraunhoffer	0.775209	−0.690953	−0.0073348	$0.02 \leq \frac{\Delta P}{P_0} \leq 1.80$
Brossard	0.515877	−0.586536	0.0436875	$0.02 \leq \frac{\Delta P}{P_0} \leq 9.00$

Table B.4 Spherical gaseous charge—scaled distance in energy versus arrival time

$\text{Ln}\,\lambda = A_0 + \sum_{i=1}^{i=n} A_i\,(\text{Ln}\,T_a.E^{-1/3})^i$				
λ (m MJ$^{-1/3}$) T_a (ms) E (MJ)				
Authors	A_0	A_1	A_2	Range $T_a.E^{-1/3}$
Prisme	−0.141626	0.61269	0.043143	$0.7 \leq T_a.E^{-1/3} \leq 16.2$
Fraunhoffer	−0.201348	0.671362	0.0327643	$3.4 \leq T_a.E^{-1/3} \leq 86.6$

Table B.5 Spherical gaseous charge—overpressure versus scaled distance in mass

$\text{Ln}\,\frac{\Delta P}{P_0} = A_0 + A_i\,(\text{Ln}\,Z)^i$				
Z (m kg$^{-1/3}$)				
Authors	A_0	A_1	A_2	Range Z
Prisme	1.88844	2.1066	0.151977	$1.6 \leq Z \leq 14.5$
Fraunhoffer	1.51497	−1.35099	−0.0283001	$1.9 \leq Z \leq 45.1$
Brossard	1.57137	−1.85455	0.119494	$1.3 \leq Z \leq 33.0$

Table B.6 Spherical gaseous charge—arrival time versus scaled distance in mass

$\text{Ln}\,T_a.M^{-1/3} = A_0 + \sum_{i=1}^{i=n} A_i\,(\text{Ln}\,Z)^i$				
T_a (ms) M (kg) Z (m kg$^{-1/3}$)				
Authors	A_0	A_1	A_2	Range Z
Prisme	−0.276392	1.73459	−0.112731	$1.6 \leq Z \leq 14.5$
Fraunhoffer	0.213769	1.40205	−0.050966	$1.9 \leq Z \leq 45.1$

Table B.7 Spherical gaseous charge—scaled distance in mass versus overpressure

$\text{Ln}\,Z = A_0 + \sum_{i=1}^{i=n} A_i\,(\text{Ln}\,\frac{\Delta P}{P_0})^i$				
Z (m kg$^{-1/3}$)				
Authors	A_0	A_1	A_2	Range $\frac{\Delta P}{P_0}$
Prisme	0.955941	−0.548635	0.0360441	$0.07 \leq \frac{\Delta P}{P_0} \leq 2.40$
Fraunhoffer	1.10782	−0.693276	−0.00646115	$0.02 \leq \frac{\Delta P}{P_0} \leq 1.8$
Brossard	0.903496	−0.57828	0.043468	$0.02 \leq \frac{\Delta P}{P_0} \leq 9.00$

Table B.8 Spherical gaseous charge—scaled distance in mass versus arrival time

$\text{Ln}\, Z = A_0 + \sum_{i=1}^{i=n} A_i\, (\text{Ln}\, T_a.M^{-1/3})^i$				
$Z\,(\text{m}\,\text{kg}^{-1/3})$ T_a (ms) M (kg)				
Authors	A_0	A_1	A_2	Range $T_a.M^{-1/3}$
Prisme	0.181225	0.546425	0.0431455	$1.7 \leq T_a.M^{-1/3} \leq 34.9$
Fraunhoffer	−0.0976254	0.655406	0.0319497	$4.6 \leq T_a.M^{-1/3} \leq 125$

Appendix C
Parameters Values for Fitted Curves: High Explosives

See Tables C.1, C.2, C.3, and C.4.

Table C.1 High explosives—overpressure versus scaled distance in mass

$\text{Ln}\,\frac{\Delta P}{P_0} = A_0 + A_i\,(\text{Ln}\,Z)^i$								
$Z\,(\text{m kg}^{-1/3})$								
Authors	Explosive	Shape	A_1	A_2	A_3	A_4	A_5	Range Z
CEA	TNT	H	2.7043	−2.04609	−0.100365			$0.7 \leq Z \leq 5.3$
ISL	TNT	C	3.60115	−3.28159	0.389002			$2.4 \leq Z \leq 15.0$
ISL	C4	H	3.27543	−2.82461	0.279555			$2.3 \leq Z \leq 11.9$
ISL	TNT	S	2.06566	−2.46315	−0.015517	0.417603	−0.146383	$0.5 \leq Z \leq 4.0$
ISL	Comp B	S	3.31713	−4.96016	2.08603	−0.451185		$1.9 \leq Z \leq 9.0$

Table C.2 High explosives—arrival time versus scaled distance in mass

$\text{Ln}\,T_a.M^{-1/3} = A_0 + \sum_{i=1}^{i=n} A_i\,(\text{Ln}\,Z)^i$						
T_a (ms) M (kg) $Z\,(\text{m kg}^{-1/3})$						
Authors	Explosive	Shape	A_1	A_2	A_3	Range Z
CEA	TNT	H	−0.706702	1.61779	0.0755105	$0.7 \leq Z \leq 5.3$
ISL	TNT	C	−0.915251	2.11672	0.172515	$2.4 \leq Z \leq 15.0$
ISL	C4	H	−1.05188	2.21441	−0.190921	$2.3 \leq Z \leq 11.9$
ISL	TNT	S	−0.481358	1.73386	0.204594	$0.5 \leq Z \leq 4.0$
				A_4	A_5	
				−0.324934	0.116912	
ISL	Comp B	S	−0.915305	2.27438	−0.235374	$1.9 \leq Z \leq 9.0$

I. Sochet (ed.), *Blast Effects*, Shock Wave and High Pressure Phenomena,
https://doi.org/10.1007/978-3-319-70831-7

Table C.3 High explosives—scaled distance in mass versus overpressure

$\mathrm{Ln}\, Z = A_0 + \sum_{i=1}^{i=n} A_i\, (\mathrm{Ln}\, \frac{\Delta P}{P_0})^i$

$Z\ (\mathrm{m\, kg^{-1/3}})$

Authors	Explosive	Shape	A_1	A_2	A_3	A_4	Range $\frac{\Delta P}{P_0}$
CEA	TNT	H	1.24724	−0.433903	−0.0101891		$0.4 \leq \frac{\Delta P}{P_0} \leq 30.0$
ISL	TNT	C	1.27801	−0.437927	0.0582008		$0.08 \leq \frac{\Delta P}{P_0} \leq 2.70$
ISL	C4	H	1.33016	−0.487331	0.0408447		$0.13 \leq \frac{\Delta P}{P_0} \leq 2.90$
ISL	TNT	S	0.93181	−0.500665	0.0195717		$0.7 \leq \frac{\Delta P}{P_0} \leq 41.0$
ISL	Comp B	S	0.996605	−0.47964	0.0723202	0.0234477	$0.1 \leq \frac{\Delta P}{P_0} \leq 2.40$

Table C.4 High explosives—scaled distance in mass versus arrival time

$\mathrm{Ln}\, Z = A_0 + \sum_{i=1}^{i=n} A_i\, (\mathrm{Ln}\, T_a.M^{-1/3})^i$

$Z\ (\mathrm{m\, kg^{-1/3}})\ T_a\ (\mathrm{ms})\ M\ (\mathrm{kg})$

Authors	Explosive	Shape	A_1	A_2	A_3	Range $T_a.M^{-1/3}$
CEA	TNT	H	0.428191	0.595771	−0.0151712	$0.3 \leq T_a.M^{-1/3} \leq 8.8$
ISL	TNT	C	0.496118	0.43566	0.0522684	$2.3 \leq T_a.M^{-1/3} \leq 35.0$
ISL	C4	H	0.524918	0.438821	0.048705	$2.07 \leq T_a.M^{-1/3} \leq 26.08$
ISL	TNT	S	0.271728	0.609692	−0.0299967	$0.2 \leq T_a.M^{-1/3} \leq 6.9$
				A_4	A_5	
				−0.0085532	0.0103069	
ISL	Comp B	S	0.443027	0.429335	0.0558045	$0.6 \leq T_a.M^{-1/3} \leq 6.4$

Further Readings

Adushkin, V. V., & Korokov, A. I. (1961). Parameters of a shock wave near to HE charge at explosion in air. *Prikladnoi Mekhaniki i Tekhnicheskoi Fiziki, 5*, 119–123. In Russian.

Akhavan, J. (2004). *The chemistry of explosive* (2nd ed.). Cambridge: Royal Society of Chemistry.

Allied Ammunition Storage and Transport Publication (AASTP)-4 (2003). Explosives Safety Risks Analysis, Part II - AC/258(ST)WP/221.

Anderson, O. L. (1984). A universal thermal equation of state. *Journal of Geodynamics, 1*, 185–214.

ASTM (2001). The ASTM computer program for chemical thermodynamic and energy release evaluation CHETAH 7.3, August 2001.

Baker, W. E. (1973). *Explosions in air*. Austin: University of Texas Press.

Baker, W. E., Cox, P. A., Westine, P. S., Kulesz, J. J., & Strehlow, R. A. (1983). *Explosion hazards and evaluation*. Fundamental Studies in Engineering. New York: Elsevier.

Banks, M., & Abernathy, R. (2004). Data-base of range evaluated improvised explosives (D-BREIE) phase III. Report No. FR=03-17, Energetic Materials Research and Testing Center, Socorro, NM (May 2004).

Batchelor, G. (1996). *The life and legacy of G. I. Taylor*. Cambridge: Cambridge University Press.

Behrens, K., & Schneider, H. (1975). Ausbreitungsfunktionen sphärischer Luftstosswellen für den Fall detonierender Äthylen-Luft-Gemische. Fraunhofer ICT.

Ben-Dor, G. (1991). *Shock wave reflection phenomena*. New York: Springer.

Bennet, F. D. (1959). Flow fields produced by exploding wires. In W. G. Moore & H. K. Chace (Eds.), *Exploding Wires Proceedings* (Vol. 1, pp. 211–266). New York: Plenum Press.

Berthelot, M. (1892). *Explosives and their power (translated from French to English by Hake and Macnab)*. London: John Murray.

Bethe, H. A., Fuchs, K., Hirschfelder, H. O., Magee, J. L., Peierls, R. E., & von Neumann, J. (1947). Blast wave. LASL 2000, Los Alamos Scientific Laboratory (distributed March 27, 1958).

Biot, J. B. (1802). Théorie mathématique de la propagation du son. *Journal of Physics, 55*, 173–182.

Bogosian, D., Ferritto, J., & Shi, Y. (2002, August 13–15). Measuring uncertainty and conservatism in simplified blast models. In *30th Explosive Safety Seminar, Atlanta, GA*.

Bogosian, D., Yokota, M., & Rigby, S. (2016, September). TNT equivalence of C-4 and PE-4: A review of traditional sources and recent data. In *MABS 24, 24th International Symposium on the Military Application of Blast and Shock, Proceedings, Halifax, Canada*.

Borenstein, E., & Benaroya, H. (2009). Sensitivity analysis of blast loading parameters and their trends as uncertainty increases. *Journal of Sound and Vibration, 321*, 762–785.

I. Sochet (ed.), *Blast Effects*, Shock Wave and High Pressure Phenomena,
https://doi.org/10.1007/978-3-319-70831-7

Borgers, J., & Vantomme, J. (2008, August 12–14). Improving the accuracy of blast parameters using a new Friedlander curvature α. Department of Defence (DOD) explosives safety seminar, Palm Springs, CA.

Borisov, A. A., Gelfand, B. E., & Tsyganov, S. A. (1985). On modelling of pressure waves formed by detonation and combustion of gas mixtures. *Fizika Goreniya i Vzryva (Novosibirsk, USSR), 21*(2), 90–97.

Brill, A., Me-Bar, Y., Sadot, O., & Ben-Dor, G. (2012). A method for measuring the impulse on structural foundations due to a blast wave. *International Journal of Impact Engineering, 49*, 214–221.

Brode, H. L. (1955). Numerical solutions of spherical blast waves. *Journal of Applied Physics, 26*(6), 766–775.

Brossard, J., Bailly, P., Desrosier, C., & Renard, J. (1988). Overpressures imposed by a blast waves. *Progess in Astronautics and Aeronautics, AIAA, Washington DC, 114*, 389–400.

Brossard, J., Leyer, J. C., Desbordes, D., Saint-Cloud, J. P., Hendricks, S., Garnier, J. L., et al. (1985). Air blast from unconfined gaseous detonation. *Progress in Astronautics and Aeronautics, AIAA, Washington DC, 94*, 556–566.

Brouillette, M. (2002). The Richmeyer-Meshkov instability. *Annual Review of Fluid Mechanics, 34*, 445–468.

Browning, S., Sherburn, A., & Schwer, E. (2013). Predicting blast loads using LSDYNA and CTH. In *Proceedings, ASCE Structures Congress, Pittsburgh.*

Buntzen, R. R. (1962). The use of exploding wires in the study of small-scale underwater explosions. In W. G. Moore & H. K. Chace (Eds.), *Exploding Wires Proceedings* (Vol. 2, pp. 195–205). New York: Plenum.

Campidelli, M., Tait, M., El-Dakhakhni, W., & Mekky, W. (2015). Inference of blast wave front parameter uncertainty for probabilistic risk assessment. *Journal of Structural Engineering, 141*(12), 04015062.

Chaloupka, J. L., Woods III, M., Aas, J., Hutchins, J., & Thistle, J. D. (2014). Color schlieren imaging with a two-path, double knife edge system. *Optics Express, 22*, 8041–8046.

Chandra, N., Ganpule, S., Kleinschmit, N. N., Feng, R., Holmberg, A. D., Sundaramurthy, A., et al. (2012). Evolution of blast wave profiles in simulated air blasts: Experiment and computational modeling. *Shock Waves, 22*(5), 403–415.

Chase, W. G., & Moore, H. K. (Eds.) (1959). *Exploding wires* (4 Vols.). New York: Plenum.

Cheval, K., Loiseau, O., & Vala. V. (2010). Laboratory scale tests for the assessment of solid explosive blast effects. Part I: Free-field test campaign. *Journal of Loss Prevention in the Process Industries, 5*, 613–621.

Chock, J., & Kapania, R. (2001). Review of two methods for calculating explosive air blast. *The Shock and Vibration Digest, 33*(2), 91–102.

Cleaver, R. P., & Robinson, C. G. (1996). An analysis of the mechanisms of overpressure generation in vapour cloud explosions. *Journal of Hazardous Materials, 45*, 27–44.

Cooper, P. W. (1994). Comments on TNT equivalence. In *20th International Pyrotechnics Seminar Colorado Springs, CO*, July 24–29, 2014.

Cooper, P. W. (1996). *Explosives engineering*. New York: Wiley-VCH.

Cranz, C. (1926). *Lehrbuch der ballistik.* Berlin: Springer.

Davis, W. C. (1997). Shock waves, rarefaction waves, equations of state. In J. A. Zukas & W. P. Walters (Eds.), *Explosive effects and applications*. Berlin: Springer.

Dharaneepathy, M., Kwshava Rao, M., & Santhakumar, A. (2006). Critical distance for blast resistant design. *Computers & Structures, 54*(4), 587–595.

Dennis, K., Maleya, L., Liang, Z., & Radulescu, M. I. (2014). Implementation of large scale shadowgraphy in hydrogen explosion phenomena. *International Journal of Hydrogen Energy, 39*(21), 1346–11353.

Desbordes, D., Manson, N., & Brossard, J. (1978). Explosion dans l'air de charges sphériques non confinées de mélanges réactifs gazeux. *Acta Astronautica, 5*(11–12), 1009–1026 (Published in French).

Desrosier, C., Reboux, A., & Brossard, J. (1991). Effect of asymmetric ignition on the vapor cloud spatial blast. *Progess in Astronautics and Aeronautics, AIAA, Washington DC, 134*, 21–37.

Dewey, J. M. (1964). The air velocity in blast waves from t.n.t. explosions. *Proceedings of the Royal Society A, 279*, 366–385.

Dewey, J. M. (1971). The properties of a blast wave obtained from an analysis of the particle trajectories. *Proceeding of the Royal Society of London A, 324*, 275–299.

Dewey, J. M. (1985). The propagation of sound from the eruption of Mount St. Helen's on 18th May 1980. *Northwest Science, 59*(2), 79–91.

Dewey, J. M. (1997a). Shock waves from explosions. In S. F. Ray (Ed.), *High speed photography and photonics* (Chap. 16). Oxford: Focal Press.

Dewey, J. M. (1997b). Explosive flows: Shock tubes and blast waves. In W.-J. Yang (Ed.), *Handbook of flow visualization* (2nd ed., Chap. 29) New York: Hemisphere.

Dewey, J. M. (2001). Spherical shock waves: Chapter 13.1 expanding spherical shocks (Blast Waves). In G. Ben-Dor, O. Igra, T. Elperin (Eds.), *Handbook of shock waves*. Amsterdam: Elsevier.

Dewey, J. M. (2005). The TNT equivalence of an optimum propane-oxygen mixture. *Journal of Physics D: Applied Physics, 38*, 4245–4251.

Dewey, J. M. (2015). Measurement of the physical properties of blast waves. In O. Igra & F. Seiler (Eds.), *Experimental methods of shock wave research* (pp. 53–86). Berlin: Springer.

Dewey, J. M. (2017). A user interface to provide the physical properties of blast waves from propane explosions. In *Proceedings of 22nd International Symposium Military Aspects of Blast and Shock, MABS22, Halifax, Canada*.

Dewey, J. M., McMillin, D. J., & Classen, D. F. (1977). Photogrammetry of spherical shocks reflected from real and ideal surfaces. *Journal of Fluid Mechanics, 81*, 701–717.

Dewey, J. M., & Sperrazza, J. (1950). The effect of atmospheric pressure and temperature on air shock. BRL Report 721, Aberdeen Proving Ground, MD.

Dewey, M. C. & Dewey, J. M. (2014). The physical properties of the blast wave produced by a stoichiometric propane/oxygen explosion. *Shock Waves, 24*, 593–601.

Dobashi, R. (2008, July 7–11). Study on consequence analyses of blast wave generated by gaseous deflagrations. In *7th International Symposium on Hazards, Prevention, and Mitigation of Industrial Explosions (ISHPMIE), St Petersburgh, Russia* (pp. 322–329).

Dobratz, B., & Crawford, P. (1985). LLNL explosives handbook. Lawrence Livermore National Laboratory, University of California, Livermore.

Dorofeev, S. B. (1995). Blast effects of confined and unconfined explosions. In *20th International Symposium on Shock Waves, Pasadena, CA, USA* (Vol. 1, pp. 77–86).

Dorofeev, S. B. (2007). Evaluation of safety distances related to unconfined hydrogen explosions. *International Journal of Hydrogen Energy, 32*, 2118–2124.

Dorofeev, S. B., Sidorov, V. P., Dvoinishnikov, A. E., Alekseev, V.I., & Kyznetsov, M. S. (1993). Experimental study of air blast parameters from fuel rich mixtures detonation. In *14th International Colloquium on Dynamics of Explosions and Reactive Systems (ICDERS), 2/E1.4.1, Coimbra, Portugal*.

Dvoinishnikov, A. E., Dorofeev, S. B., & Gelfand, B. E. (1995). Analysis of blast wave data from HE explosions. In *Shock waves* (pp. 407–412). Berlin, Heidelberg: Springer.

Earnshaw, S. (1858). On the mathematical theory of sound. Report of the 28th Meeting British Association (pp. 34–35).

Earnshaw, S. (1860a). On the velocity of the sound of thunder. *Philosophical Magazine, 20*(IV), 37–41.

Earnshaw, S. (1860b). On the mathematical theory of sound. *Philosophical Transactions of the Royal Society of London, 150*, 133–148.

Eichinger, W. E. (1985). Mach stem modeling with spherical shock waves. Thesis Faculty of the School of Engineering of the Air Force Institute of Technology Air University.

Emanuel, G. (2000). *Analytical fluid dynamics* (2nd ed.). Boca Raton, FL: CRC.

Emanuel, G. (2001). Theory of shock waves. In G. Ben Dor, O. Igra, & T. Elperin (Eds.), *Handbook of shock waves* (Vol. 1, Chap. 3.1), Theoretical, Experimental, and Numerical Techniques. New York: Academic.

Enstock, L. K., & Smith, P. D. (2007). Measurement of impulse from the close-in explosion of doped charges using a pendulum. *International Journal of Impact Engineering, 34*, 487–494.

Esparza, E. D. (1986). Blast measurements and equivalency for spherical charges at small scaled distances. *International Journal of Impact Engineering, 4*(1), 23–40.

Ethridge, N. H. (1978). Proposed design for a differential pressure gage to measure dynamic pressure in blast wave. Defense Nuclear Agency.

Filler, W. S. (1956). Post-detonation and thermal studies of solid high explosives in a closed chamber. *Combustion of Explosives and Solid Propellants, 6*, 648–657.

Fishburn, B. D. (1976). Some aspects of blast from fuel-air explosives. *Acta Astronautica, 3*, 1049.

Formby, S. A., & Wharton, R. K. (1996). Blast characteristics and TNT equivalence values for some commercial explosives detonated at ground level. *Journal of Hazardous Materials, 50*, 183–198.

Fouchier, C., Laboureur, D., Youinou, L., Lapebie, E., & Buchlin, J. M. (2017). Experimental investigation of blast wave propagation in an urban environment. *Journal of Loss Prevention in the Process Industries*. https://doi.org/10.1016/j.jlp.2017.06.021

Friedlander, F. G. (1946). The diffraction of sound pulses. I. Diffraction by a semi-infinite plate. *Proceedings of the Royal Society of London A, 186*, 322–344.

Gault, K. (2017). Ondes de choc en milieu confiné. Private report. DGATN (in French).

Gelfand, B. E. (2004). Translation from Russian to English the Book "Blast effects caused by explosions" authored by B. Gelfand and M. Silnikov. United States Army, European Research Office of the U. S. Army, London, England, Contract Number N62558-04-M-0004.

Gelfand, B. E., Gubin, C. A., Mikhalkin, V. N., & Shargatov, V. A. (1985). Computation of shock wave parameters by detonation of combustible gaseous mixtures of a variable composition. *Journal of Physics of Combustion and Explosion (Russian Academy of Sciences), 3*, 92.

Gitterman, Y. (2014). Secondary shock features for large surface explosions: Results from the Sayarim Military Range, Israel and other experiments. *Shock Waves, 24*, 267–282.

Glass, I. I. (1958). Shock tubes. Part 1: Theory and performance of simple shock tubes. UTIA Rev. No. 12, Pt 1, University of Toronto Institute for Aerophysics (now University of Toronto Institute for Aerospace Studies), Toronto, Canada.

Goel, M., Matsagar, V., Gupta, A., & Marburg, S. (2012). An abridged review of blast wave parameters. *Defence Science Journal, 62*(5), 300–306.

Goodman, H. J. (1960). Compiled free-air blast data on bare spherical pentolite. AD No. 235275. Aberdeen Proving Ground, MD: U. S. Army Ballistic Research Laboratory.

Grimwall, G. (1999). *Thermophysical properties of materials*, Enlarged and Revised Edition. Amsterdam: North-Holland, Elsevier.

Hargather, M. J., & Settles, G. S. (2007). Optical measurement and scaling of blasts from gram-range explosive charges. *Shock Waves, 17*, 215–223.

Hargather, M. J., & Settles, G. S. (2009). Retroreflective shadowgraph technique for large-scale flow visualization. *Applied Optics, 48*(22), 4449–4457.

Held, M. (1999). Impulse method for the blast contour of cylindrical high explosive charges. *Propellants, Explosives, Pyrotechnics, 24*, 17–26.

Held, M. (2001). Improved momentum method. *Propellants, Explosives, Pyrotechnics, 26*, 290–295.

Held, M., Jager, E. H., & Stolzl, B. (1961). TNT-blast equivalence forbursting or pressurized gas conventional vessels. Paper at 6th SMIRT Conference, Paris.

Henrich, J. (1979). *The dynamics of explosions*. Amsterdam: Elsevier.

Higashino, F., Henderson, L. F., & Shimizu, F. (1991). Experiments on the interaction of a pair of cylindrical weak blast waves in air. *Shock Waves, 1*, 275–284.

Hokanson, J. C., Esparza, E. D., Baker, W. E., Sandoval, N. R., & Anderson, C. E. (1982). Determination of blast loads in the DWF, I, II, SwRI - 6578. San Antonio, TX.

Hopkinson, B. (1915). British Ordnance Board Minutes, 13565.

Hugoniot, P. H. (1887). Mémoire sur la propagation du mouvement dans les corps et ples spécialement dans les gaz parfaits, Partie 1. *Journal de l'École Polytechnique (Paris), 57*, 3–97 (published in French).
Hugoniot, P. H. (1889). Mémoire sur la propagation du mouvement dans les corps et plus spécialement dans les gaz parfaits, Partie 2. *Journal de l'École Polytechnique (Paris), 58*, 1–125 (published in French).
Hulton, F. G., & Enstock, L. (2000). Inert matter in close-in explosive loading: An experimental investigation using a pendulum. In *MABS16, 16th International Symposium Military Aspects of Blast and Shock*, RMCS, Cranfield University (pp. 365–370).
International Ammunition Technical Guideline (IATG) (2011). Formulae for ammunition management 01.80, United Nations.
Ismail, M. M., & Murray, S. G. (1993). Study of the blast wave parameters from small scale explosion. *Propellants, Explosives, Pyrotechnics, 18*, 11–17.
Jeremie, R., & Bajie, Z. (2006). An approach to determining the TNT equivalent of high explosives. *Scientific and Technical Review, 56*(1), 58–62.
Jiang, Z., Takayama, K., Moosad, K. P. B., Onodera, O., & Sun, M. (1998). Numerical and experimental study of a micro-blast wave generated by pulsed-laser beam focusing. *Shock Waves, 8*, 337–349.
Julien, B., Sochet, I., & Vaillant, T. (2016). Impact of the volume of rooms on shock wave propagation within a multi-chamber system. *Shock Waves, 26*(2), 87–108.
Karlos, V., & Solomos, G. (2016). Analysis of blast parameters in the near-field for spherical free-air explosions. JRC Technical Report, EUR 27823EN, December. Luxembourg: European Union.
Karlos, V., Solomos, G., & Larcher, M. (2016). Analysis of the blast wave decay coefficient using the Kingery–Bulmash data. *International Journal of Protective Structures, 7*(3), 409–429.
Kingery, C. N. & Bulmash, G. (1984). Airblast parameters from TNT spherical air burst and hemispherical surface burst. US Technical Report ARBRL-TR-02555. Ballistics Research Laboratory, Aberdeen Proving Ground, Maryland, USA (April 1984).
Kinney, G. F., & Graham, K. J. (1985). *Explosives shocks in air* (2nd edn.). Berlin: Springer.
Kisters, T., Kuder, J., & Nau, S. (2016). Autonomous gauge for blast impulse determination close to explosive charges. *Shock Waves, 26*, 117–127.
Kleine, H. (2001). Measurements techniques and diagnostics. In G. Ben Dor, O. Igra, & T. Elperin (Eds.). *Handbook of shock waves* (Vol. 1, Chap. 5.1). Theoretical, Experimental, and Numerical Techniques. New York: Academic.
Kleine, H., & Takayama, K. (2002). Visualization of laboratory-scale blast wave phenomena. In M. Kawahashi (Ed.). *10th International Symposium on Flow Visualization*, paper F0149, Kyoto, Japan.
Kleine, H., & Takayama, K. (2004). Laboratory-scale blast wave phenomena. In B. Milton, T. Saito, & M. Sun (Eds.). *Proceedings of the Symposium on Interdisciplinary Shock Wave Research*, Tohoku University (pp. 257–276).
Kleine, H., Dewey, J. M., Oashi, K., Mizukai, T., & Takayama, K. (2003). Studies of the TNT equivalence of silver azide charges. *Shock Waves, 13*, 123–138.
Kleine, H., Timofeev, E., & Takayama, K. (2005). Laboratory scale blast wave phenomena – optical diagnostics and applications. *Shock Waves, 14*(5/6), 343–357.
Klomfass, A. (2010). Investigations on fluid dynamic instabilities and pressure fluctuations in the near field of detonations. In *MABS 21, 21st International Symposium on the Military Application of Blast Shock, Jerusalem, Israel.*
Kogarko, S. M., Adushkin, V. V., & Liamin, A. G. (1965). Investigation of spherical detonations in gas mixtures. *Combustion, Explosion and Shock Waves, USSR, 1*, 2–22.
Kondrikov, B. N., & Sumin, A. I. (1987). Equation of state for gases at high pressure. *Combustion, Explosion, and Shock Waves, 23*, 105–113.
Krauthammer, T. (2008). *Modern protective structures*. Boca Raton, FL: CRC, Taylor &Francis Group.

Krehl, P. (2001). History of shock waves. In G. Ben Dor, O. Igra, & T. Elperin (Eds.), *Handbook of shock waves* (Vol. 1, Chap. 1). Theoretical, Experimental, and Numerical Techniques. New York: Academic.

Lannoy, A. (1984). Analyse des explosions air-hydrocarbure en milieu libre: Etudes déterministes et probabiliste du scénario d'accident. Prévision des effets de surpression, Bulletin Direct. Etudes et Recherche EDF. A4 (published in French).

Larcher, M. (2007). Simulation of the effects of an air blast wave. JRC Technical Report, EUR 41337EN, January. Luxembourg: European Union.

Lee, E., Horning, J., & Kury, J. (1968). Adiabatic expansion of high explosives detonation products. Lawrence Livermore National Laboratory, University of California, Livermore, TID 4500-UCRL 50422.

Lees, F. P. (1996). *Loss prevention in the process industries, hazard identification, assessment and control* (Vol. 2, 2nd ed.). Oxford: Butterworth Heinemann.

Lefranļois, A., Baudin, G., Cremoux, J. L., Massoni, J., & Saurel, R. (2002). Blast efficiency of aluminized high explosives. In *MABS17, 17th International Symposium on the Military Application of Blast and Shock, Las Vegas, Nevada, USA*.

Lide, D. R. (Ed.) (2004). *Handbook of chemistry and physics* (84th ed.). Boca Raton, FL: CRC.

Liepmann, H. W., & Roshko, A. (1957). *Elements of gas dynamics*. Galcit Aeronautical Series. New York: Wiley.

Lin, S. C. (1954). Cylindrical shock waves produced by instantaneous energy release. *Journal of Applied Physics, 25*, 54–57.

Lu, J. P. (2001). Evaluation of the Thermochemical Code CHEETAH 2.0 for Modelling Explosives Performance, DSTO-TR-1199.

Mach, E., & Sommer, J. (1877). Uber die Fortpflanzunggeshwindigkeit von explosionsschall-wellen. Akademie der Wissenschaften, Sitzangberichte der Wiener, 74.

Mader, L. (1963). Detonation properties of condensed explosives computed using the Becker-Kistiakosky-Wilson equation of state. Los Alamos Scientific Laboratory Report LA-2900, New Mexico.

Maffeo, M., Carboni, M., Cyganik, J., Decristofano, B., Carneal, C., Zinn, D., et al. (2016). Test method development for the evaluation of head borne equipment with a blast simulator. In *MABS 24, 24th International Symposium on Military Aspect of Blast and Shock, Halifax, Canada*.

Maillot, Y., Sochet, I., Vinļont, J. Y., Grillon, Y. (2017). Etude expérimentale de la réflexion de Mach, 23ème Congrès Franļais de mécanique, Lille, France (Published in French).

McGee, B. C., Hobbs, M. L., & Baer, M. R. (1998). Exponential 6 Parameterization of the JCZ3-EOS", Sandia Report SAND98-1191. http://prod.sandia.gov/techlib/access-control.cgi/1998/981191.pdf

McNesby, K. L., Biss, M. M., Benjamin, R. A., & Thompson, R. A. (2014). Optical Measurement of peak air shock pressures following explosions. *Propellants, Explosives, Pyrotechnics, 39*, 59–64.

Merzkirch, W. (1987). *Flow visualization* (2nd ed.). New York: Academic.

Meshkov, E. E. (1970). Instability of a shock wave accelerated interface between two gases. Izv. Akad. Nauk. SSSR, Mekh Zhidk. i. Gaz, 151–158 (NASA translation TT F-13 R. F.).

Meyer, R., Köhler, J., & Homburg, A. (2007). In R. Meyer, J. Köhler, & A. Homburg (Eds.), *Explosives* (6th ed.). Weinheim: Wiley-VCH Verlag GmbH.

Mills, C. A. (1987). The design of concrete structure to resist explosions and weapon effects. In *Proceedings of the 1st International Conference on Concrete for Hazard Protections, Edinburgh, UK* (pp. 61–73).

Miura, A., Matsuo, A., Mizukaki, T., Shiraishi, T., Utsunomiya, G., Takayama, K., et al. (2004). Reflection and diffraction phenomena of blast wave propagation in nuclear fuel cycle facility. *Japan Society Mechanical Engineering, 47*(2), 287–292.

Mizukaki, T., Kleine, H., & Takayama, K. (2001). Visualization of blast waves in the early stage of milligram charge explosions. In *Proceedings of 18th International Colloquium on Dynamics of Explosions and Reactive Systems (ICDERS), Seattle, USA, paper 171*.

Muthurajan, H., Sivabalan, R., Talawar, M. B., Venugopalan, S., & Gandhe, B. R. (2006). Computer code for the optimization of performance parameters of mixed explosive formulations. *Journal of Hazardous Materials A, 136*, 475–481.

Nansteel, M. W., Veldman, R. L., Chen, C. C. T., & Lawrence, W. (2013). Impulse plug measurements of blast reflected impulse at close range. *Propellants, Explosives, Pyrotechnics, 38*, 120–128.

NASA Lewis Research Center. (2000). http://www.me.berkeley.edu/gri_mech/data/thermo_table.html

Naz, P. (2005). Etude de souffle des têtes explosives. Etude bibliographique. Institut Franco-Allemend de Recherche de Saint-Louis - R 129/2005 (in French).

Needham, C. E. (2010). *Blast waves*. Berlin: Springer.

Neuberger, A., Peles, S., & Rittel, D. (2007). Scaling the response of circular plates subjected to large and close-range spherical explosions. Part I: Air-blast loading. *International Journal of Impact Engineering, 34*, 859–873.

Neuwald, P., & Reichenbach, H. (2002). Detonations in front of a tunnel en-trance: a parametric small-scale study. In *MABS 17; 17th International Symposium on Military Aspect of Blast and Shock, Las Vegas, Nevada, USA*.

Ngo, T., Lumantarna, R., & Whittaker, A. (2015). Quantification of the blast-loading parameters of large-scale explosions. *Journal of Structural Engineering: American Society of Civil Engineers, 141*, 04015009.

Ngo, T., Mendis, P., Gupta, A., & Ramsay, J. (2007). Blast loading and blast effects on structures - An overview. *Electronic Journal of Structural Engineering Special Issue: Loading on Structures, 7*, 76–91.

Ohashi, K., Kleine, H., & Takayama, K. (2002). Characteristics of blast waves generated by milligram charges. In F. Lu (Ed.), *23rd International Symposium on Shock Waves, Fort Worth, USA* (pp. 187–193).

Omang, M., Christensen, S. O., Borve, S., & Trulsen, J. (2009). Height of burst explosions: A comparative study of numerical and experimental results. *Shock Waves, 19*(2), 135–143.

Ornellas, D. L. (1982). Calorimetric determinations of the heat and products of detonation for explosives: October 1961 to April 1982. Technical Report UCRL-52821, Lawrence Livermore Laboratory, University of California.

Ornellas, J. D. (1968). Heat and products of detonation of cyclotetramethylenetetranitramine, 2,4,6-trinitrotoluene, nitromethane, and bis[2,2-dinitro-2-fluoroethyl]formal. *Journal of Physical Chemistry, 72*, 2390–2394.

Ostraich, B., Kivity, Y., Anteby, I., Sadot, O., & Ben-Dor, G. (2009). Load assesment on safe rooms doors report. Shock Tube Laboratory. Protective Technologies R&D Center, Faculty of Engineering Sciences, Ben-Gurion University of the Negev, Beer-Sheva, Israel (in Hebrew).

Parmentier, G. (1993). Synthèse des résultats expérimentaux relatifs aux détonations d'explosifs sphériques. Institut Franco-Allemand de Recherche de Saint-Louis, R 113/93, Contrat no. 91.02.178/ETBS-CETAM 18 Juin 1993 (published in French).

Peiris, S. M., & Gump, J. G. (2008). Static compression of energetic materials. In S. M. Peiris & G. J. Piermarini (Eds.), *Equations of state and high pressure phases for explosives*. Berlin: Springer.

Peugeot, F., Deschalbault, E., & Péron, P. F. (2006, October). TNT equivalency: Misconceptions and reality. Munitions Safety Information Analysis Center, MSIAC Unclassified, L-132.

Pförtner, H. (1977). Gas cloud explosions and resulting blast effects. *Nuclear Engineering and Design, 41*, 59–67.

Piehler, T., Birk, A., Benjamin, R., Boyle, V., Summers, E., & Aubert, S. (2009). Near-field impulse loading measurement techniques for evaluating explosive blast. Army Research Laboratory, ARL-RP-235.

Pierorazio, J. A., Thomas, J. K., Baker, Q. A., & Ketchum, D. E. (2005). An update to the Baker-Strelhow-Tang vapor cloud explosion prediction methodology flame speed table. *Process Safety Progress, 24*(1), 59–65.

Poisson, S. D. (1808). Mémoire sur la théorie du son, *Journal de l'École Polytechnique (Paris), 7*, 319–392 (published in French).
Poling, B. E., Pausnitz, J. M., & O'Connell, J. P. (2001). *The properties of gases and liquids* (5th edn., Chap. 5). Boston: McGraw-Hill.
Prandtl, L., & Tietjens, O. G. (1957). *Fundamentals of hydro- and aerodynamics*. Engineering Societies Monographs. New York, NY: Dover. Reprinted by Dover. 1934, First publication.
Protective Design Center, United States Army Corps of Engineers (2007, May 22). CONWEP, Conventional Weapons Effects, https://pdc.usace.army.mil/software/conwep/
Rae, P. J., & Gunderson, J. (2016). Characterization of a large shock tube. In *MABS24, 24th International Symposium on Military Aspect of Blast and Shock, Halifax, Canada*.
Ram, O., & Sadot, O. (2012). Implementation of the exploding wire technique to study blast-wave–structure interaction. *Experiments in Fluids, 53*, 1335–1345. https://doi.org/10.1007/s00348-012-1339-8
Rankine, W. J. M. (1870a). On the thermodynamic theory of waves of finite longitudinal disturbance (read 16 Dec., 1869). *Philosophical Transactions of the Royal Society of London, 160*, 277–286.
Rankine, W. J. M. (1870b). supplement to "On the thermodynamic theory of waves of finite longitudinal disturbance". *Philosophical Transactions of the Royal Society of London, 160*, 287–288.
Reichenbach, H., & Neuwald, P. (2000). Fluid-dynamics of explosions in multi-chamber systems phenomenology test program. Ernst-Mach Institute, Freiburg, Germany.
Reichenbach, H., & Neuwald, P. (2001). Indoor detonations - visualization and pressure measurement in small-scale models. In *24th International Congress on High-Speed Photography and Photonics*. Proceedings of SPIE (Vol. 4183, pp. 92–104). ISBN 9780819438461.
Reichenbach, H., Neuwald, P., & Kuhl, A. L. (2002). Role of precision laboratory experiments in the understanding of large-scale blast phenomena. In *MABS17: Julius J. Meszaras Lecture, 17th International Symposium on Military Aspect of Blast and Shock, Las Vegas, Nevada, USA*.
Reisler, R. E., Keefer, J. H., & Ethridge, N. H. (1995). MABS Monograph, air blast instrumentation, 1943–1993, Measurement Techniques and Instrumentation, 2, The High Explosive Era, 1959–1993 - Defense Nuclear Agency.
Reithel, R. J., Blackburn, J. H., Seay, GE., & Skolnick, S. (1962). The current pause in an exploding wire. In W.G. Moore & H. K. Chace (Eds.), *Exploding wires proceedings* (Vol. 1, pp. 19–32). New York: Plenum.
Remennikov, A. M. (2007). The state of the art of explosive loads characterisation, University of Wollongong, Wollongong, Australia, available online at http://www.aees.org.au
Richtmyer, R. D. (1960). Taylor instability in shock acceleration of compressible fluids. *Communications on Pure and Applied Mathematics, 13*, 297–319.
Rigby, S. E., & Sielicki, P. W. (2014b). An investigation of TNT equivalence of hemispherical PE4 charges. *Engineering Transactions, 62*(4), 423–435.
Rigby, S. E., Tyas, A., Bennett, T., Clarke, S. D., & Fay, S. D. (2014a). The negative phase of the blast load. *International Journal of Protective Structures, 5*(1), 1–20.
Rigby, S. E., Tyas, A., Clarke, S., Fay, S.D., Reay, J.J., Warren, J.A., Gant, M., & Elgy, I. (2015). Observations from preliminary experiments on spatial and temporalpressure measurements from near-field free air explosions. *International Journal of Protective Structures, 6*(2), 175–190.
Ripley, R. C., Dunbar, T. E., Donhaue, L., & Von Rosen, B. (2004). Personnel vulnerability predictions using small-scale air blast modeling. In *MABS18, 18th International Symposium on Military Aspect of Blast and Shock, Bad Reichenhall, Germany*.
Ripley, R. C., Von Rosen, B., Ritzel, D. V., & Whitehouse, D. R. (2004). Small-scale modeling of explosive blasts in urban scenarios. In *21st International Symposium on Ballistics, Adelaide, Australia*.
Ritzel, D. V., & Parks, S. A. (2010). Blast simulation using shock-tube technology. In *MABS21, 21st International Symposium on Military Aspect of Blast and Shock, Jerusalem, Israel*.

Robey, R. (2001). Blast tubes. In G. Ben-Dor, O. Igra, T. Elperin (Eds.), *Handbook of shock waves* (Vol. 1, Chap. 4.4). New York: Academic.

Rose, T. A., & Smith, P. D. (2002). Influence of the principal geometrical parameters of straight city streets on positive and negative phase blast wave impulses. *International Journal of Impact Engineering, 27*, 239–376.

Rupert, V. (1991). Shock-interface interaction: Current research on the Richtmyer-Meshkov problem. In K. Takayama (Ed.), *18th International Symposium on Shock Waves* (Vol. 1, pp. 83–94).

Sachs, R. G. (1944). The dependence of blast and ambient pressure and temperature. BRL Report 466.

Sakurai, A. (1965). Blast wave theory. In M. Holt (Ed.), *Basic developments in fluid mechanics* (Vol. 1, pp. 309–375). New York: Academic.

Saska, P., Krzysta, E., & Mezyk, A. (2011). An analysis of an explosive shock wave impact on to military vehicles of contemporary warfare. *Journal of KONES Powertrain and Transport, 18*, 1.

Schultz, E., & Shepherd, J. (2000). Validation of detailed reaction mechanisms for detonation simulation. Explosion Dynamics Laboratory Report FM99-5. http://authors.library.caltech.edu/25820/1/FM99-5.pdf

Scilly, N. F. (1995). Measurement of the explosive performance of high explosives. *Journal of Loss Prevention in the Process Industries, 8*(5), 265–273.

Settles, G. S. (2001). *Schlieren and shadowgraph techniques*. Heidelberg, New York: Springer.

Shin, J., Whittaker, A., & Cormie, D. (2015a). Incident and normally reflected overpressure and impulse for detonations of spherical high explosives in free air. *Journal of Structural Engineering, 41*, 12.

Shin, J., Whittaker, A., Cormie, D.,& Willford, M. (2015b). Design charts and polynomials for air-blast parameters. In M. G. Stewart & M. D. Netherton (Eds.), *3rd International Conference on Protective Structures (ICPS3), Newcastle, Australia*, 3–6 February 2015.

Slater, J. E., Boechler, D. E., & Edgar, R. C. (1995). DRES Measurement of free-field airblast. Minor Uncle Symposium Report. Defense Nuclear Agency, POR 7453-4, 4, 2, 1–98.

Smith, P. D., & Hetherington, J. G. (1994). *Blast and ballistic loading of structures* (pp. 145–223). Oxford: Butterworth-Heinemann.

Smith, P. D., Mays, G. C., Rose, T. A., Teo, K. G., & Roberts, B. J. (1992). Small scale models of complex geometry for blast over pressure assessment. *International Journal of Impact Engineering, 12*, 345–360.

Smith, P. D., & Rose, T. A. (2000). Influence of urban geometry on blast wave resultants. In *MABS16, 16th International Symposium on Military Aspect of Blast and Shock, Oxford, England*.

Smith, P. D., Rose, T. A., & Ng, S. H. (2004). The influence of areal density on the shielding and channeling of blast by buildings. In *MABS18, 18th International Symposium on Military Aspect of Blast and Shock, Bad Reichenhall, Germany*.

Smith, P. D., Vismeg, P., Teo, L. C., & Tingey, L. (1998). Blast wave transmission along rough-walled tunnels. *International Journal of Impact Engineering, 21*(6), 419–432.

Sochet, I., Eveillard, S., Vinļont, J. Y., Piserchia, P. F., & Rocourt, X. (2017). Influence of the geometry of protective barriers on the propagation of shock waves. *Shock Waves, 27*(2), 209–219.

Sochet, I., Sauvan, P. E., Boulanger, R., & Nozeres, F. (2014a). Effect of a gas charge explosion at the closed end of a gas storage system. *Journal of Loss Prevention in the Process Industries, 27*, 42–48.

Sochet, I., Sauvan, P. E., Boulanger, R., & Nozeres, F. (2014b). External explosion in an industrial site. *Journal of Loss Prevention in the Process Industries, 29*, 56–64.

Sochet, I., & Schneider, H. (2010). Blast wave characteristics and equivalency. In S. M. Frolov, F. Zhang, & P. Wiolanski (Eds.), *Explosion, dynamics hazards* (pp. 169–184). Moscow: Torus Press.

Stokes, G. G. (1848). On a difficulty in the theory of sound. *Philosophical Magazine, 34*(III), 349–356.

Strehlow, R. A. (1975). Blast waves generated by constant velocity flames: A simplified approach. *Combustion and Flame, 24*, 257–261.

Swisdak, M. M. J. (1975). Explosion effects and properties- Part I - Explosion effects in air. Report NSWC/WOL/TR-116.

Swisdak, M. M. J. (1994). Simplified kingery airblast calculations. In *Proceedings of the 26th Department of defense (DoD) Explosives Safety Seminar, Miami, FL.*

Takayama, K., Sasoh, A., Onodera, O., Yang, J., Ojima, H., Ogawa, T., et al. (1994). Holographic interferometric observation of weak shock waves generated by entrance of a high-speed train into a long tunnel. In J.-S. Chang & S-H. Park (Eds.), *Proceedings of 21st International Congress High-Speed Photo. Photonics*, SPIE (Vol. 2513, pp. 294–303).

Tanaka, K. (1985). Detonation properties of high explosives calculated by revised Kihara–Hikita equation of state. In *Proceedings 8th Symposium (International) on Detonation, Albuquerque/New Mexico/Washington, DC.*

Tang, M. J., & Baker, Q. A. (1999). A new set of blast curves from vapor cloud explosion. *Process Safety Progress, 18*(3), 235–240.

Taylor, G. I. (1946). The air wave surrounding an expanding sphere. *Proceedings of the Royal Society A, 186*, 273–292.

Taylor, G. I. (1950a). The instability of liquid surfaces when accelerated in a direction perpendicular to their plane. *Proceedings of the Royal Society A, 210*, 192–196.

Taylor, G. I. (1950b). The formation of a blast wave by a very intense explosion. I. Theoretical discussion. *Proceeding of the Royal Society of London Series A, Mathematical and Physical Sciences, 201*(1065), 159–174.

Thibault, P. (2009). Review of equation of state models. Chemical Equilibrium Calculations and CERV Code Requirements for SHS Detonation Modelling. Contract Defence R&D Canada, Report DRDC Suffield CR 2010–013.

Thornhill, C. K. (1957). The ultimate distribution of energy in a spherical explosion. Armament Research and Development Establishment (ARDE) Report (B) 27/57, HMSO, London.

Thornhill, C. K. (1959). The shape of a spherical blast wave. Armament Research and Development Establishment (ARDE) Memo. (B) 41/59, HMSO, London.

Todoroff, V. (2013). Mesure d'un champ de masse volumique par background oriented schlieren 3D. Etude d'un dispositif expérimental et des méthodes de traitement pour la résolution expérimental et des méthodes de traitement pour la résolution du problème inverse. Thèse de doctorat de l'université de Toulouse (09 Décembre 2013) (published in French).

Tongchang, Y., Menchao, Y., & Jianling, W. (1995). Determination of heats of detonation and influence of components of composite explosives on heats of detonation of high explosives. *Journal of Thermal Analysis, 44*, 1347–1356.

Trélat, S., Sochet, I., Autrusson, B., Cheval, K., & Loiseau, O. (2007b). Impact of a shock wave on a structure on explosion at altitude. *Journal of Loss Prevention in the Process Industries, 20*, 509–516.

Trélat, S., Sochet, I., Autrusson, B., Cheval, K., & Loiseau, O. (2007a). Strong explosion near a parallelepipedic structure. *Shock Waves, 16*(4–5), 349–357.

Trzcinski, W. A. (2015) On some methods of determination of the detonation energy of explosives. https://www.researchgate.net/publication/266224730

UN SaferGuard User Survey (2015). International Ammunition Technical Guideline IATG 0.180, Formulae for ammunition management, Second edition 2015-02-01.

U. S. Army Corps of Engineers, Naval Facilities Engineering Command, Air Force Civil Engineer Support Agency (2008). Technical Manuals, Unified Facilities Criteria (UFC), Structures to resist the effects of accidental explosions, UFC 3-340-02.

U. S. Department of the Army, the Navy and the Air Force (1990). Design of structures to resist the effects of accidental explosions, Technical Manual TM5-1300, Washington DC.

U. S. Department of Energy (1980). A Manual for Prediction of Blast and Fragment Loadings on Structures. DOE/TIC - 11268.

Van den Berg, A. C. (1985). The multi-energy method: A framework for vapor cloud explosion blast prediction. *Journal of Hazardous Materials, 12*(1), 1–10.

Van Wylen, G. J., & Sonntag, R. E. (1985). *Fundamentals of classic thermodynamics* New York: Wiley.

Wharton, R.K, Formby, S. A., & Merrifield, R. (2000). Airblast TNT equivalence for a range of commercial blasting explosives. *Journal of Hazardous Materials, A79*, 31–39.

Wong, F. C. H., Gottlieb, J.J., & Lussier, L. S.: Chemical Equilibrium analysis of combustion products at constant volume. Defence R&D, Canada - Valcartier, Technical Report, DRDC Valcartier TR 2003-375, 2003-12-22.

Zhang, F., Yoshinaka, A., Anderson, J., & Ripley, R. (2006). Confined heterogeneous blast. In *MABS19, 19th International Symposium on Military Aspects of Blast and Shock*, Def. R. & D. Suffield, AB, Canada.

Zhdan, S. A. (1983). Calculation of gas mixture explosion with regard to shift of chemical equilibrium products. *Journal of Physics of Combustion and Explosion (Russian Academy of Sciences), 1*, 131.

Zyskowski, A., Sochet, I., Mavrot, G., Bailly, P., & Renard, J. (2004). Study of the explosion process in a small scale experiment – structural loading. *Journal of Loss Prevention, 17*, 291–299.

Index

A

Arrival time, 4, 8, 12, 18, 34, 57–71, 91, 95, 97–99, 103, 104, 106, 113, 116–119, 131, 134, 135, 137

B

Blast wave, 1, 3, 7, 17–35, 37–45, 47–54, 57–59, 141–168
- characteristics, 98–109
- definition, 1
- effects, 5
- energy, 5
- measurement, 11–12
- properties, 3–4, 38, 138
- source, 1–2, 5, 7
- structures, 5, 9–10

BLEVE, 7

C

Chapman-Jouguet, 95
Charge voltage, 74
Cylindrical blast, 144, 151–156

D

Decomposition, 123–125
- Kistiakowsky–Wilson, 123
- products, 123
- Springall–Roberts, 124

Deflagration, 1, 3, 5, 7
Density, 18, 19, 21, 22, 25, 27, 33–35
Detonation, 1, 3, 6–8, 90, 91, 94, 98
- energy, 76, 77, 122, 125–126, 138
- products, 2, 6, 107, 122, 123

Duration, 3, 4, 8, 38, 42–46, 51, 91, 97, 98, 105–107
Dynamic pressure, 11

E

Energy, 5, 142, 144, 146, 148, 149, 151, 156, 158, 159, 163, 165
- mechanical, 2
- released, 2, 5–7, 94, 96–98

Enthalpy, 20, 94, 125–126
Entropy, 3, 5, 18, 32–33
Expansion, 2
Experiment(s), 141–168
- large scale, 91, 97–98, 103–107, 141, 143
- medium scale, 89, 95–97, 103–107, 114–119
- shock tube, 89
- small scale, 89, 90, 92–93, 95, 103–107, 114–119, 141–168

Exploding wire, 73–76, 78, 92, 94, 141–168
- charge voltage, 76–78, 92
- copper, 75, 76, 92
- diameter, 76, 78
- energy, 76, 77
- high voltage, 144
- length, 76–78
- zinc-copper, 58, 92

Explosion, 3, 57, 60, 66
- air burst, 116

I. Sochet (ed.), *Blast Effects*, Shock Wave and High Pressure Phenomena, https://doi.org/10.1007/978-3-319-70831-7

Explosion (*cont.*)
chemical, 2
free-air burst, 91, 114–116
nuclear, 5
source, 5
surface, 116
surface burst, 91, 114–116
thermodynamics, 122–129
unconfined, 2

F
Friedlander, 37–54, 89

G
Gaseous charge, 94–95
Gaseous explosion, 59, 65, 66, 91, 93
ethylene-air, 95, 97
hemispherical charge, 91, 93, 95–107
propane-oxygen, 48, 52, 94, 97
spherical charge, 91, 93, 96–97, 99–107
Gaseous explosion(s), 134–136, 138

H
Height-of-burst, 10–11, 99, 116–119
High explosive, 1, 2, 7, 89, 90, 135–138, 144
ANFO, 8, 41, 45, 53, 54
C4, 6, 114–119
composition B, 114–119, 127–129
cylindrical charge, 114–119
hemispherical charge, 114, 116–119
RDX, 6, 127–129
spherical charge, 114, 116–119
thermobaric, 7
TNT, 6, 8, 53, 54, 90, 99, 113–119, 143, 148, 165–167
Hopkinson-Cranz, 98, 156, 165, 167
Hydrostatic pressure, 3, 4, 8, 9, 11

I
Impulse, 3, 4, 8, 12, 42, 44, 46, 47, 91, 96–98, 103–107, 116–119, 139, 143, 156–159, 161–168
pendulum, 12
Initiation, 8, 77, 93, 95

M
Mach number, 12, 13, 18–23, 25, 27–29, 31, 33, 34, 58, 64, 68, 76, 131, 156
Measure, 3, 6, 11, 18, 25, 34, 35, 58, 63, 65, 66, 70, 97

O
Oxygen balance, 123

P
Particle velocity, 18, 19, 23, 25, 40, 41, 46
Plasma, 74, 75
Power discharge, 74
Pressure, 142, 143, 145, 147–150, 152, 153, 155–168
dynamic, 17, 25–26, 33, 34, 40, 41, 46, 58
hydrostatic, 17, 19, 21–22, 29, 38–41, 49–52, 59, 62, 65, 66, 68, 69
over-, 17, 19, 21, 26–31, 33, 34, 38, 40–52, 91, 96–99, 103–107, 116–119, 134–139
profile, 78, 79, 86, 99, 105, 108
reflected, 18, 30, 149, 159, 162
sensor, 58
total, 17, 23, 27, 29, 40, 53, 57
Pressure sensor(s), 58, 147
electronic, 11
passive, 11
piezoelectric, 38, 40, 43, 58, 93, 96, 97, 99, 114, 147
piezoelectric ceramic, 58, 59, 63, 65, 66, 68–71
Pitot, 39, 40, 58

R
Rankine-Hugoniot, 13, 19, 58, 62, 67, 98

S
Scale, 7–8
Scaled distance, 156
energy, 98, 103–107
mass, 98, 103–107, 116–119
Scaling laws, 8
Scaling laws for blast wave
Hopkinson-Cranz, 8–9
Shock, 3
primary, 3, 6, 8, 11, 12, 58, 69
secondary, 3, 12
Shock reflection, 69
Shock wave, 57, 75–78, 89, 91, 99
first, 76, 77, 86
primary, 108
primary shock, 17–19, 26, 29, 34, 35

reflected, 29–31, 116–119
reflection, 99, 105
second, 76, 77, 86
secondary, 108–109
Shock wave primary, 60–65, 108
Sound speed, 17, 21, 23–25, 31–34

T
Temperature, 18, 20, 21, 24, 32–34
reflected, 31–32
TNT, 6
TNT equivalence, 6, 129–139
arrival time, 131
Chapman-Jouguet, 131
explosion yield, 131–133
impulse, 130
multi-parameter, 133
pressure, 130

V
Visualisation, 35, 74, 76, 77, 99
high-speed, 12, 73, 97
photography, 35, 53, 141–143, 152
schlieren, 142, 143
smoke tracer, 12, 53
Voltage signal, 62, 69

The manufacturer's authorised representative in the EU is Springer Nature Customer Service Centre GmbH, Europaplatz 3, 69115 Heidelberg, Germany. If you have any concerns regarding our products, please contact ProductSafety@springernature.com

Printed and bound by CPI Group (UK) Ltd, Croydon, CR0 4YY
25/09/2026
02222836-0003